T0291042

Mathematical Modelling of Heat Transfer Performance of Heat Exchanger using Nanofluids

The book presents a detailed discussion of nanomaterials, nanofluids and application of nanofluids as a coolant to reduce heat transfer. It presents a detailed approach to the formulation of mathematical modelling applicable to any type of case study with a validation approach and sensitivity and optimization.

- Covers the aspects of formulation of mathematical modelling with optimization and sensitivity analysis.
- Presents a case study based on heat transfer improvement and performs operations using nanofluids.
- Examines the analysis of experimental data by the formulation of a mathematical model and correlation between input data and output data.
- Illustrates heat transfer improvement of heat exchangers using nanofluids through the mathematical modelling approach.
- Discusses applications of nanofluids in cooling systems.

This book discusses the aspect of formulation of mathematical modelling with optimization and sensitivity analysis. It further presents a case study based on the heat transfer improvement and performing operations using nanofluids. The text covers sensitivity analysis and analysis from the indices of the model. It also discusses important concepts such as nanomaterials, applications of nanomaterials, and nanofluids. It will serve as an ideal reference text for senior undergraduate and graduate students in fields including mechanical engineering, chemical engineering, aerospace engineering, industrial engineering, and manufacturing engineering.

Mathematical Modelling of Heat Transfer Performance of Heat Exchanger using Nanofluids

Prashant Maheshwary,
Chandrahas C. Handa,
Neetu Gyanchandani and Pramod Belkhode

CRC Press
Taylor & Francis Group
Boca Raton London New York

CRC Press is an imprint of the
Taylor & Francis Group, an **informa** business

First edition published 2024
by CRC Press
2385 Executive Center Drive, Suite 320, Boca Raton, FL 33431

and by CRC Press
4 Park Square, Milton Park, Abingdon, Oxon, OX14 4RN

CRC Press is an imprint of Taylor & Francis Group, LLC

© 2024 Prashant Maheshwary, Chandrahas C. Handa, Neetu Gyanchandani and Pramod Belkhode

Library of Congress Cataloging-in-Publication Data
Names: Maheshwary, Prashant, author. | Handa, Chandrahas C, author. | Gyanchandani, Neetu, author. | Belkhode, Pramod, author.
Title: Mathematical modelling of heat transfer performance of heat exchanger using nanofluids / by Prashant Maheshwary, Chandrahas C. Handa, Neetu Gyanchandani, Pramod Belkhode.
Description: First edition. | Boca Raton, FL : CRC Press, 2024. | Includes bibliographical references and index.
Identifiers: LCCN 2023010739 (print) | LCCN 2023010740 (ebook) | ISBN 9781032478753 (hbk) | ISBN 9781032557656 (pbk) | ISBN 9781003432111 (ebk)
Subjects: LCSH: Heat exchangers—Mathematical models. | Working fluids. | Nanofluids. | Heat—Transmission—Mathematical models.
Classification: LCC TJ263 .M34 2024 (print) | LCC TJ263 (ebook) | DDC 621.402/5—dc23/eng/20230331
LC record available at https://lccn.loc.gov/2023010739
LC ebook record available at https://lccn.loc.gov/2023010740

ISBN: 978-1-03247-875-3 (hbk)
ISBN: 978-1-03255-765-6 (pbk)
ISBN: 978-1-00343-211-1 (ebk)

DOI: 10.1201/9781003432111

Typeset in Sabon
by Apex CoVantage, LLC

Contents

Preface

Nanoscience and nanotechnology in engineering directly link to industries and daily life. As a result, plentiful nanomaterials, nanodevices and nanosystems for diverse engineering purposes have been developed and used for the betterment of human beings. In a similar manner, nanofluids also find a wide range of applications in thermal engineering. Nanofluid is an adulterate suspension of nanoparticles dispersed in a liquid. Among various studies carried out in the assessment of heat transfer performance of heat exchangers using nanofluids, no researcher has applied the "Theories of Engineering Experimentation" as suggested by Hilbert (1967), which quantitatively specifies the influence of an individual independent variable on a response variable. A theoretical approach can be adopted wherein a known logic can be applied correlating the various dependent and independent parameters of the system. Though qualitatively, the relationships between the dependent and independent parameters are known based on the available literature references, the generalized quantitative relationships are sometimes not known. Hence, formulating the quantitative relationship based on the known logic is not possible. On account of there being no possibility of formulation of a theoretical model, one is left with the only alternative of formulating an experimental data-based model. Once the model is formulated for any phenomenon, one gets a clear idea about the variation of dependent variables in terms of their interaction with various independent variables. Then, applying the optimization techniques, determining the optimum conditions for the execution of this phenomenon becomes possible. Hence, any experimental research should ultimately get precipitated to formulation of a model. Systematically applying this approach, the model gives a new insight into optimization of parameters for each industrial activity separately in order to achieve maximum productivity.

With heat transfer performance of a heat exchanger using nanofluids, it is only partially possible to plan experimentation. However, in many such systems, test planning part of the experimentation approach is not feasible to be adopted. One has to allow the activity (i.e., phenomenon) to take place either the way it takes place or else allow it to take place as planned by others. This happens when one wishes to formulate a model for heat transfer performance of a heat exchanger using nanofluids. The approach adopted for formulating a generalized experimental data-based model suggested by Hilbert Schenck Jr. has been proposed in this work, involving steps such as identification of variables or parameters affecting the phenomenon, reduction of variables through dimensional analysis, test planning, rejection of absurd data and formulation of the model.

Identification of dependent and independent variables of a phenomenon is to be done based on known qualitative physics of the phenomenon. If the system involves a large number of independent variables, the experiment becomes tedious, time-consuming and costly. By deducing dimensional equation for a phenomenon, we can reduce the number of independent variables. The exact mathematical form of this equation will be the targeted model. Upon getting the experimental results, adopting the appropriate method for test data checking and rejection, the erroneous data are identified and removed from the gathered data. Based on this purified data, one has to formulate a quantitative relationship between the dependent and independent dimensionless terms of the dimensional equation. The main objective of the present work is to establish a quantitative relationship of interaction of inputs on response variables as well as to optimize them in the heat transfer performance of a heat exchanger using nanofluids.

We are thankful to those who have helped us directly and indirectly towards the successful completion of this work.

—**Authors**

Author Bios

Dr. Prashant Maheshwary is Dean, Faculty of Science and Technology, Rashtrasant Tukadoji Maharaj Nagpur University, Nagpur. He did his double doctoral research degree in two different specializations (machine design and thermal engineering) of mechanical engineering. His area of research is nanomaterials and nanoscience, thermal engineering and mechanisms and machines. Subjects taught earlier are heat transfer, thermodynamics, characterization of nanomaterials and application of nanomaterials at UG and PG levels. An accomplished teaching professional with more than three decades of teaching experience and 10 years of research experience, he has published more than 20 research papers in SCI and Scopus (A+)-indexed international journals. In-depth administrative experience gained being the Director of an Educational Institute since 2008. He is a self-driven, result-oriented person with flexibility and ability to connect to all levels in an organization. Continuous self-development and continuous learning have added to his knowledge and is an asset to the institution he is associated with.

Dr. Chandrahas C. Handa is Professor and Head of the Department of Mechanical Engineering at Karmaveer Dadasaheb Kannamwar College of Engineering (KDKCE), Rashtrasant Tukadoji Maharaj Nagpur University (RTMNU), Nagpur, India. His area of research includes photovoltaic cells and nano sensors. He teaches subjects such as machine design, optimization techniques and genetic logarithm. He obtained his PhD in mechanical engineering from RTMNU, Nagpur. He has more than 33 years of teaching and administrative experience and 25 years of research experience. He has published over 120 research papers in international/national journals and presented 99 papers in international/national conferences. He has guided 12 PhD students at RTMNU, Nagpur. He is the recipient of five Best Teacher Awards, including one such award given by RTMNU, Nagpur. He has received many research, development and training grants of more than Rs. 2 Crore from various Central Ministries of Government of India. He has published four patents and five copyrights. He is the former Treasurer of ISTE, New Delhi. He has worked as Principal, Vice Principal, Dean of

Engineering College in the past. He has developed more than 50 machines and received research grants from AICTE and various industries.

Dr. (Mrs.) Neetu Gyanchandani did her doctoral research in the area of electronics engineering. She is currently Dean (R & D), S. B. Jain Institute of Technology, Management and Research, Nagpur, Maharashtra, India. Her area of research is the application of nanospintronics, composite materials and image processing. Subjects she teaches include application of nanospintronics, computer materials and image processing at UG and PG levels. She is a capable academician and an administrator with more than 18 years of teaching experience. She has been awarded as best women teacher of Engineering College by Indian Society of Technical Education, New Delhi. She is also the single point of contact (sPoC) for indigenous MOOCs platform Swayam NPTEL, since 2018. Under her leadership, a center of excellence in industrial robotics was started by IIT Mumbai in her department. She has published more than five patents and ten research papers.

Dr. Pramod Belkhode did his doctoral research degree in mechanical engineering from Rashtrasant Tukadoji Maharaj Nagpur University, Nagpur. He has been Assistant Professor at Laxminarayan Institute of Technology since 2009. His area of research is mathematical modelling and simulation. He has published more than 50 research papers in international journals, secured four patents and delivered more than 30 guest lectures on various topics.

Chapter 1

Nanofluids

1.1 NANOTECHNOLOGY

Nanotechnology is an area of technology based on nanoscience. Besides the technological relevance of nanoscience, there is enormous excitement associated with nanotechnology. Fantastic claims have been made about faster computers, cheap production of goods and medical breakthroughs. Nanotechnology is expected to appear in products such as tennis rackets, self-cleaning cars, paint, food, cosmetics and thermal underwear. Governments are investing billions of dollars into nanoscience research. The enhancement in thermal conductivity is still a thrust area of research, which is not fully explored by researchers, scientists and academicians [1]. In 2004, the worldwide investment in nanotechnology research and development was estimated to be $8.6 billion. The American nanotechnology effort is called the National Nanotechnology Initiative [2]. The European Union has identified nanotechnology as an important research area and has spent €1300 million on nanotechnology research in the period 2002–2006 [3].

1.2 NANOMATERIALS

A reduction in the spatial dimension or confinement of particles or quasi-particles in a particular crystallographic direction within a structure generally leads to changes in physical properties of the system in that direction [4]. Hence, another classification of nanostructured materials and systems essentially depends on the number of dimensions which lie within the nanometer range: (a) 3D systems are confined in three dimensions, e.g. structures typically composed of consolidated equiaxed crystallites; (b) 2D systems are confined in two dimensions, e.g. filamentary structures where the length is substantially greater than the cross-sectional dimensions; (c) 1D systems are confined in one dimension, e.g. layered or laminate structures; (d) 0D or zero-dimensional structures are nanopores and nanoparticles [5, 6].

DOI: 10.1201/9781003432111-1

1.3 APPLICATION OF NANOMATERIALS

Nano-sized materials currently are used in numerous industries, e.g., carbon black particles make rubber tires wear-resistant; nanofibers are used for insulation and reinforcement of composites; iron oxide creates the magnetic material used in disk drives and audio-video tapes; nano-zinc oxides and titania are used as sun blocks for UV rays; etc. Nanoscale particles and nanothin layers of materials are being used among other things to make products lighter, stronger or more conductive. Some of the products available in the market using nanotechnology include magnetic recording tapes, computer hard disks, bumpers on cars, solid-state compasses, protective and glare-reducing coatings for eyeglasses and windows, automobile catalytic converters, metal cutting tools, dental bonding agents, longer-lasting tennis balls, burn and wound dressing, ink, etc. One of the important applications in nanomedicine is the creation of nanoscale devices for improved therapy and diagnostics. Such nanoscale devices or nanorobots serve as vehicles for delivery of therapeutic agents, detectors or guardians against early disease and perhaps repair of metabolic or genetic defects. For applications in medicine, the major challenge is "miniaturization": new instruments to analyze tissues literally down to the molecular level, sensors smaller than a cell allowing to look at ongoing functions and small machines that literally move within a human body pursuing pathogens and neutralizing chemical toxins [7].

1.4 NANOFLUIDS

Nanofluids, on the other hand, offer many advantages over the single phase pure fluids and suspensions with micro particle. The problems of particle sedimentation due to gravity, clogging of micro channel passage and erosion of tube material are minimized to a great extent when nanofluids are used in heat exchangers. Besides, nanofluids form stable suspensions with uniform dispersion of nano particles in the base fluid. Thermo-physical properties of single phase heat transfer fluids such as water, oils and glycols are well established and are available in literature and hand books. The thermo-physical properties of two phase nanofluids are not explored much. Nanofluids are considered to be an alternate and new generation liquids for transport of heat energy and can be employed as heat transfer fluids in heat exchangers in place of pure single phase fluids. The applications of nanofluid in heat transfer include radiators in automotives, chemical engineering and process industries, solar water heater, refrigeration, cooling of electronics devices. The main objective of obtaining heat transfer enhancement using nanofluids is to accommodate high heat fluxes and hence to reduce the cost and size of the heat exchangers which in turn results in conservation of energy and material.

1.5 COMPACT HEAT EXCHANGERS

The importance of compact heat exchangers (CHEs) has been recognized in aerospace, automobile, gas turbine power plant and other industries for the last 50 years or more. This is due to several factors such as packaging constraints, sometimes high performance requirements, low cost and use of air or gas as one of the fluids in the exchanger. For nearly two decades the additional driving factors for heat exchangers design have been reducing energy consumption for operation of heat exchangers and process plants, and minimizing the capital investment [8–10]. Due to efficient heat transfer characteristics of nanofluids, fabrication of compact heat exchangers becomes possible.

1.6 HEAT TRANSFER ENHANCEMENT THROUGH NANOFLUIDS

Nanofluids have great potential to enhance the heat transfer characteristics of base fluid through improved thermo-physical properties. In this section, a brief overview has been presented to know recent development in the field of heat transfer enhancement using nanofluids. During literature survey, it is observed that heat transfer enhancement is function of several parameters such as concentration, particle size, shape of nanoparticles, temperature, viscosity of nanofluid, Brownian motion etc.

Ismael et al. recognized in their work, the addition of highly conductive particles can significantly increase the thermal conductivity of heat transfer fluids [11]. Sridhara et al. comprehensively summarized the recent development in Al_2O_3-based nanofluids. This review focused on stability of nanofluids, enhancement of thermal conductivity, viscosity and heat transfer characteristics of Al_2O_3-based nanofluids. The review's recommendation is that Al_2O_3 nanoparticles with particle size ranging from 13 to 302 nm shows enhancement in the thermal conductivity to the tune of 2% to 36% [12].

Wang et al. broadly studied the fluid flow and heat transfer characteristics of nanofluids in forced and convection flow and identified the potential applications of nanofluids for near future research [13]. According to kinetic theory, the kinetics of aggregates depends on random motion of nanoparticles and induced micro-convection. Nabi et al. proposed that stability of nanofluids play a crucial role in the enhancement of thermal conductivity. This work concludes that more the nanofluid is stabilized; the more effective is the thermal conductivity [14]. Kleinstreuer et al. comments on the lack of consistency in experimental data and measurement methods for thermal conductivity measurements. These works also question whether the anomalous enhancement in thermal conductivity of nanofluid is real or takes place only due to temperature, nanoparticle size/shape and aggregation state. The conclusion of said work is that clear benchmark experimentation is required to understand the proper mechanism involved in the enhancement of thermal conductivity [15].

Goharshadi *et al.* shed light on several other parameters which affects the thermal conductivity of nanofluids, such as shape, size, concentration of nanoparticles, temperature and pH [16]. Nsofor *et al.* demonstrated that nanoparticles in liquid suspensions have superior thermal conductivities to those of the base fluids [17]. Liu *et al.* shows that beside concentration, size and temperature, stability of nanofluid also has significant effect on thermal conductivity. The good stability of nanofluid can be achieved by proper use of surfactants or by adding the proper ammonia content in the fluid [18]. Hong *et al.* enlisted the parameters such as particle size, effect of surfactant and dispersion of particles, which influence the thermal properties of nanofluids. In this report, thermal properties of Fe and TiO_2 nanofluids were studied as a function of volume fractions [19].

Timofeeva *et al.* developed the novel approach to reduce viscosity of alumina nanofluid without significantly affecting thermal conductivity. This report underlines that surface charge of nanoparticles also have an important role in viscosity, and it is controlled by adjusting the pH of nanofluid [20]. Wong *et al.* identified that agglomeration and clustering are great problems that affect the performance of nanofluids. Therefore, focused research efforts are needed to minimize this unwanted clustering [21]. Cabaleiro *et al.* prepared the ZnO/glycol water nanofluids and optimized stability characteristics. This study was carried out using Dynamic Light Scattering. Good agreement is established between experimental data expressed through dimensionless Nusselt number [22].

1.7 IMPROVEMENT IN HEAT EXCHANGER PERFORMANCE

This section summarizes the important published work on the enhancement of convection heat transfer in heat exchangers using nanofluids on two topics. It also focuses on presenting the experimental results for performance of heat exchangers. The effect of use on nanofluids on the various types of heat exchangers such as plate heat exchangers, shell and tube heat exchangers, compact heat exchangers and double pipe heat exchangers are also discussed.

Huminic *et al.* discussed in detail the application of nanofluids in different types of heat exchangers such as plate heat exchangers, shell and tube heat exchangers, compact heat exchangers and double pipe heat exchangers [23]. Mohammadian *et al.* modelled two micro-pin-fin heat exchangers and used cold and hot surfaces of a thermoelectric module to analyze the effects of Al_2O_3 water nanofluids [24]. Albadr *et al.* used Al_2O_3 nanoparticles of about 30 nm diameter to prepare nanofluids. The results of this study show that convective heat transfer coefficient of nanofluid is higher over the base liquid at constant mass flow rate and at the same inlet temperature. Also, increasing concentration of Al_2O_3 nanoparticles in base fluid causes an increase in the viscosity and friction factor [25]. Shahrul et al. indicated the improvement in the performance of the shell and tube heat exchanger by about 35% using ZnO–W nanofluids [26]. Wu et al. concludes that the increase in viscosity of the nanofluids is much larger than the thermal conductivity enhancement. This statement was confirmed using experimentation on carbon nanotube/water nanofluid, the results of which indicate that thermal conductivity increases only 1.04 times while the increase in relative viscosity is 9.56 for a 1.0 wt% [27]. Albadr et al. demonstrated that the heat transfer coefficient of the nanofluid increases with an increase in the mass flow rate, and that the heat transfer coefficient increases with the increase in volume concentration of the Al_2O_3 in nanofluid. But increasing the volume concentration results in increased viscosity of the nanofluid responsible for the increase in the friction factor [25].

Huang et al. proposed a new heat transfer correlation based on experimental data of water to predict the experimental data of nanofluids for thermal conductivity and viscosity. Similarly, correlation for friction factor was obtained, and it fit with the experimental data. The proposed new heat transfer correlation shows good agreement with experimental data [28]. Tiwari et al. demonstrated that use of proper nanomaterial is more predominant than that of heat transfer fluid temperature or volume flow rate. In this work, CeO_2/water, Al_2O_3/water, TiO_2/water and SiO_2/water nanofluids have the optimum volume concentrations as 0.75%, 1%, 0.75% and 1.25%, respectively, resulting in maximum heat transfer enhancements of about 35.9%, 26.3%, 24.1% and 13.9%, respectively [29]. Khairul et al. studied the heat transfer coefficients of CuO/water nanofluids based on a coiled heat exchanger, which shows enhancement of 5.90–14.24% as a function of volume concentration. The main accomplishment of this work is decreased friction factor with the increase in volume flow rate and volume concentration [30]. Garoosi et al. shows that at low Rayleigh numbers, the particle distribution is fairly nonuniform, while at high Rayleigh numbers, particle distribution remains almost uniform. This study also concludes that an optimal volume fraction of the nanoparticles at each Rayleigh number has maximum heat transfer rate in the heat exchanger [31]. The results obtained by Aly et al. showed a different behavior depending on the parameter selected for the comparison with the base fluid. When compared at the same Re or Dn, the heat transfer coefficient is increased by increasing the coil diameter and nanoparticles' volume concentration [32]. Ho et al. prepared the Al_2O_3-water nanofluids with mass fractions in the range of 0.1–1 wt.%. The results of the study clearly indicate that Al_2O_3-water nanofluid can noticeably enhance the heat-transfer performance of the natural circulation loop considered [33]. Srinivas et al. reported the enhancement in thermal conductivity of Al_2O_3, CuO and TiO_2/water nanofluids of 30.37%, 32.7% and 26.8% respectively. This work also concluded that higher values of nanofluid concentration and shell-side fluid temperature resulted in good performance of the heat exchanger [34].

Abed *et al.* demonstrated that when nanofluids are used in a forced convection, a 10% increase in average Nusselt number is observed for nanoparticles with a diameter of 20 nm. This type of corrugated channel can improve the thermal performance of heat exchangers due to compact size [35]. Sheikholeslami *et al.* studied a double-pipe heat exchanger in turbulent flow mode to analyze heat transfer in air to water. The study was conducted in the range of flow rate of water from 120 to 200 (L/h), with the temperature of water in the upper tank ranging from 70° to 90°C [36]. Goodarzi *et al.* studied the hermos-physical properties of carbon nanotube–based nanofluid. The experimental result shows that increasing the Reynolds number, Peclet number or fraction of nanomaterial improves the heat transfer characteristics of the nanofluid. The performance of the plate heat exchanger is also improved by using carbon nanotube/water as the working fluid [37].

Huminic *et al.* demonstrated that the use of nanofluids in a helically coiled tube-in-tube heat exchanger improves the heat transfer performance. The study was conducted using CuO and TiO_2 nanofluids. The maximum effectiveness was 91% observed for CuO nanofluids [38]. Teng *et al.* prepared the multiwalled carbon nanotubes–water nanofluids by a two-step method. The maximum enhancements of multiwalled carbon nanotubes–water nanofluids at 0.25 wt.% for the heat exchange capacity and efficiency factor was found to be 7.77% and 7.53%, respectively [39]. Baker *et al.* reported the improved efficiency of heat exchange systems for liquefied natural gas processing by using nanofluids. The main target of this study is to characterize graphene and Al_2O_3 nanofluids and to measure thermo-physical properties and stability. This work concluded that no significant change in the heat exchanger size was seen using nanofluid thermal properties due to the lubrication oil having 90% of the resistance to heat transfer [40].

1.8 APPLICATION OF NANOFLUID IN COOLING SYSTEMS

This section summarizes the recent reports on application of nanofluids in cooling systems. Most of the researchers agree that the optimum performance of cooling system can be achieved with a low-volume fraction of nanoparticles. But there are some problems and challenges regarding the mechanisms of cooling systems based on nanofluids. This area of research is still at its initial stage and needs further development.

Wang *et al.* studied the thermal energy storage characteristics of Cu-H_2O nanofluids as phase changes material for cooling systems. A mechanism involved in thermal energy storage characteristic was analyzed by the contact angle and thermal conductivity of nanofluids. For 0.1 wt% Cu nanoparticles, the total freezing time was reduced by 19.2%, which shows that

Cu-H_2O nanofluids have potential to be used in thermal energy storage applications [41]. Ahammed *et al.* studied the thermoelectric cooling performance of electronic devices with nanofluid in a multiport mini channel heat exchanger. The local Nusselt number showed enhancement of 19.22% and 23.92% for 0.1% and 0.2%, respectively, when compared to water when the Reynolds number is 1000 and with an input power of 400 W [42].

Mohammadian *et al.* numerically analyzed the performance of single-phase heat transfer and pressure drop of Al_2O_3-water nanofluids in micro-pin-fin heat exchangers. In this work, the effect of nanofluids' volume fraction and nanoparticles diameter on coefficient of performance (COP) and total entropy generation were studied [24]. Hasan *et al.* investigated the performance of a counterflow microchannel heat exchanger with a nanofluid as a cooling medium. Two types of nanofluids, Cu water and Al_2O_3 water, are used for this investigation. Use of nanofluids responsible for increase the effectiveness and cooling performance as a coolant due to ultra fine particles and small volume fraction [43]. Mehta *et al.* developed the lattice Boltzmann model for single-phase fluids by coupling the density and temperature distribution functions. This work simply concludes that the use of nanoparticles dispersed in deionized water enhances the heat transfer rate [44].

1.9 MATHEMATICAL MODELLING

Khoshvaght-Aliabadi *et al.* extensively studied the performance of a plate-fin heat exchanger with vortex-generator channels, two operating factors such as Reynolds number and nanoparticles concentration and seven geometrical parameters such as wing height, wing width, channel length, longitudinal wings pitch, transverse wings pitch, wings attach angle and wings attack angle. The mix model demonstrated a superior prediction of nanofluids flow inside the tested vortex-generator channel at the studied range [11]. Tohidi *et al.* studied the two passive techniques to investigate the heat transfer improvement (i.e., chaotic advection and nanofluids) in coiled heat exchangers. In this work, two different coils (one with normal configuration and another with chaotic configuration) are numerically analyzed and compared for both water and nanofluid. Results show that centrifugal force is responsible for generation of a pair of vortices called Dean-roll-cells, which effectively trapped fluid particles and caused drop in heat transfer and decrease in radial direction due to poor mixing [12].

Mohammadian *et al.* numerically analyzed the performance of single-phase heat transfer and pressure drop of Al_2O_3-water nanofluids in micro-pin-fin heat exchangers. In this work, the effect of nanofluids' volume fraction and nanoparticles' diameter on coefficient of performance and total entropy generation were studied [24]. Garoosi *et al.* surmised that there is an optimal volume fraction of the nanoparticles at each Richardson number for which the maximum heat transfer rate can be obtained. Moreover, it is found that for a constant surface area of the HAC at the entire range of the Richardson number, the rate of heat transfer is increased by changing the orientation of the HAC from horizontal to vertical. Results also indicate that at low Ri, the distribution of the solid particles remains almost uniform [31]. The results obtained by Aly et al. showed a different behavior depending on the parameter selected for the comparison with the base fluid. When compared at the same Re or Dn, the heat transfer coefficient increases by increasing the coil diameter and nanoparticles volume concentration [32]. Ebrahimnia-Bajestan et al. numerically analyzed the common two-phase model in order to obtain more accurate predictions of the heat transfer characteristics of nanofluids. This modified model investigated the effects of particle concentration, particle diameter and particle and base-fluid material on the heat transfer rate at different Reynolds numbers. The results indicated that the convective heat transfer coefficient increases with an increase in nanoparticle concentration and flow Reynolds number, while particle size has an inverse effect [45]. Ndoye et al. also showed numerically that nanofluids have great potential to improve cold chain efficiency by reducing energy consumption, emissions and global warming impact [46].

Kalyanmoy Deb and Singiresu. S. Rao [47] provide the methods of engineering optimization, its theory and applications. Miller Irwin et al. [48] provides information about an approach to reliability of models, equations of calculating reliability of model for parallel and series system. R. Ganeshan explained sample design and sampling, method of optimization techniques in his book on research methodology, Research Methodology for Engineers.

Hilbert Schenck Jr. provided insight about the theory of engineering experimentation and its actual implementation in engineering problems. He described the method for the formulation of experimental data-based models in his book on research methodology, Theories of Engineering Experimentation.

Chapter 2

Concept of Experimental Data-Based Modelling

2.1 INTRODUCTION

Any base liquid/fluid which contains nanoparticles in suspended state is known as nanofluid. Nanofluids are two phase fluids of solid-liquid mixture and are considered to be new-generation heat transfer fluids. In the recent past nanofluids have emerged as promising thermo fluids for heat transfer applications. The thermal conductivity property of nanofluids is expected to be higher than that of the base liquids. The practice of adding micron-size particles in the traditional heat transfer fluids was in existence since the time of Hamilton and Crosser.

Research techniques/methods are used to perform the research processes by the researchers. These include various methods of data collection, statistical techniques used for data analysis, i.e., correlation and regression methods, and methods used for assessment of the accuracy of the results obtained. Heat transfer performance is a highly complex and difficult process, and understanding such a phenomenon cannot be logic based. Hence it is necessary to apply a methodology of experimentation to such a process for formulating an experimental data-based model. In this chapter, the theory of experimentation by H. Schenck Jr. will be applied.

2.2 NANOFLUID FOR HEAT TRANSFER

Nanofluids offer many advantages over single-phase pure fluids and suspensions with micro particles. The problems of particle sedimentation due to gravity, clogging of micro channel passage and erosion of tube material are minimized to a great extent when nanofluids are used in heat exchangers. Besides, nanofluids form stable suspensions with uniform dispersion of nanoparticles in the base fluid. Thermo-physical properties of single-phase heat transfer fluids such as water, oils and glycols are well established and are available in literature and handbooks. The thermo-physical properties of two phase nanofluids are not explored much. An accurate measurement of properties of nanofluids is a prerequisite for determining the heat transfer coefficient of nanofluids.

The suitability of a particular nanofluid in a heat transfer application is then judged based on its heat transfer performance. Nanofluids are considered to be alternative and new-generation liquids for transport of heat energy and can be employed as heat transfer fluids in heat exchangers in place of pure single-phase fluids. The applications of nanofluids in heat transfer include radiators in automotives, chemical engineering and process industries, solar water heaters, refrigeration and cooling of electronics devices. The main objective of obtaining heat transfer enhancement using nanofluids is to accommodate high heat fluxes and hence to reduce the cost and size of the heat exchangers, which in turn conserves energy and material.

2.3 BRIEF METHODOLOGY OF THEORY OF EXPERIMENTATION

The approach of methodology of experimentation proposed by Hilbert Schenck Jr. has been used for heat transfer performance in a heat exchanger, as the nature of the phenomenon is complex. The basic approach is included in the following steps:

1. Identification of independent and dependent variables.
2. Reduction of independent variables adopting a dimensional analysis.
3. Test planning, comprising of determination of test envelope, test points, test sequence and experimentation plan.
4. Physical design of an experimental setup.
5. Execution of experimentation.
6. Purification of experimentation data.
7. Formulation of the model.
8. Model optimization.
9. Reliability of the model.
10. ANN simulation of the experimental data.

Formulation of mathematical model is done through the following approach:

1. *Identification of independent and dependent variables*:
    i. *Independent variables* affect, influence, or cause (or possibly cause) the outcome.
    ii. *Dependent variables* are influenced by or depend upon the independent variables.
    iii. *Extraneous variables* change in a random or uncontrolled manner.

2. *Reduction of variables by using dimensional analysis*: Dimensional analysis is done by using Buckingham's pi theorem. It helps to reduce the number of variables by formulating them into nondimensional groups or ratios called pi terms. Each pi term can be a group of one or more independent variables. It provides an ease in experimental planning and representation of results.

3. *Test planning*: Test planning consists of determining the test envelope, test points, test sequence and experimental plan.

   i. The *test envelope* is used to decide the range of variation of an individual independent pi term.

  ii. The *test points* are the discrete values of independent variables within the test envelope at which an experiment is conducted.

 iii. The *test sequence* is the sequence in which the test points are to be set during experimentation.

4. *Plan of experimentation*: In experiments involving two or more controlled and variable factors, extraneous variables may also be present. Such experiments, called multifactor experiments, can be planned and executed by either a classical plan or a factorial plan of experimentation. In a classical plan, only one independent variable is varied over its planned range and all other independent variables are kept at a predefined fixed value, i.e., kept constant. In a factorial plan, multiple independent variables can be varied over their planned range, while all other independent variables are kept constant. This plan is shorter to execute, more accurate for short experimental tests but it has very less applicability.

5. *Physical design of an experimental setup*: This necessary step includes deciding on specifications, procuring instrumentation and, subsequently, fabricating the setup.

6. *Execute experimentation*: Next step would be to execute experimentation as per test planning and gather data regarding causes (Inputs) and effects (Responses)

7. *Purification of data*: The next step is to purify the data using statistical methods.

8. *Comparison of developed model*: The final step is to establish the comparison and relationship between outputs (effects) and inputs (causes) using various graphs.

2.4 METHODS OF EXPERIMENTATION

The objective of planning the experimentation is to obtain reliable and accurate results with the execution of minimum number of trials. Various methods of experimentation is developed to satisfy these objectives. Theories of experimentation which are predominantly used in planning experimentations in engineering field are as follows:

1. *Taguchi method of experimentation*: The Taguchi method establishes the relation between performance and controllable variables with minimum effort. The planning/design is robust, which means that the performance is least affected by variation in noise variables. This method can be used with advantage for planning experimentation for a technically established process.
2. *Factorial experimentation*: Factorial experimentation arrives at satisfactory results with reduced effort. This method is used with advantage for planning experimentation for a technically established process.
3. *Classical plan of experimentation*: The classical plan requires maximum effort to get reliable results, but this plan is generally applied when the process is technically unknown and sufficiently complex to develop an analytical model. In this case generally, the interrelationship between the effect of interdependence of controllable variables and noise variables on performance is not known.

This chapter has attempted to develop a practical model to find the heat transfer performance in a heat exchanger. This model tries to establish the correlation between the influencing/independent factors (X) and performing/dependent factors (Y). The various elements of performance included in the model are discussed and their inclusion justified on the basis of existing research and result analysis. The correlation between (X) and (Y) can mathematically be expressed as follows:

$Y = f(X)$

Heat Transfer Function = f(Independent Thermo-Physical Variables)

$Y (Y_1/Y_2/Y_3/Y_4/Y_5/Y_6) = f(X_1, X_2, X_3, X_4, X_5, X_6)$

Y = Dependent variable or performing factors, such as:

Y_1 = Thermal conductivity based on concentration

Y_2 = Thermal conductivity based on size

Y_3 = Thermal conductivity based on shape

Y_4 = Temperature difference

Y_5 = Total heat flow

Y_6 = Heat transfer coefficient

X = Independent variables or influencing factors, such as:

X_1 = Density

X_2 = Temperature

X_3 = Concentration

X_4 = Size
X_5 = Shape
X_6 = Mass flow rate

The model presented in this chapter serves to establish an empirical relationship between dependent and independent variables of heat transfer performance in a heat exchanger by using theories of experimentation.

Chapter 3

Design of Experimentation

3.1 INTRODUCTION

This chapter aims to identify the important influencing factors and formulate a mathematical model to correlate the performance of a heat Exchanger (Y) with its influencing factors (X). This is done with the help of two experimental setups: (1) the two-wire method (TWM), and (2) the radiator as a heat engineer. The thermal conductivity of coolant used in a heat exchanger has become a very significant parameter in optimizing heat dissipation. Faster heat dissipation will result in improved efficiency of the system. The main purpose is to improve the performance of the heat exchanger by using nanofluid in place of traditional coolant. This will increase the rate of heat dissipation but also will make the heat exchangers compact by using less nanofluid as a coolant. The power used to circulate coolant around the tubes will also diminish as the quantity is reduced and the size of the heat exchanger is reduced. The general objective is to identify the significant influencing factors and formulate the model to correlate the performance of heat transfer in a heat exchanger with its influencing factors. Heat transfer performance in a heat exchanger is measured in terms of thermal conductivity, total heat flow, heat transfer coefficient and temperature difference at inlet and outlet, among other factors. Influencing factors were identified by an exhaustive literature survey, hypothesis of experimentation and discussions with experts. Statistical tools were used to predict and authenticate the significance of the factors. These factors were further used to formulate the model to predict heat transfer performance of a heat exchanger. Nano fluids play a very important role in various modern applications like automobile/industrial cooling, plate heat exchangers, shell and tube heat exchangers and compact heat exchangers. Nanofluid in the laboratory is prepared by one of two methods, namely the one-step method and the two-step method. The methodology of experimentation approach proposed by Hilbert Schenck Jr. has been used for the heat transfer process in a heat exchanger, as the nature of the phenomenon is complex.

DOI: 10.1201/9781003432111-3

3.2 DESIGN OF EXPERIMENTATION – METHODICAL APPROACH

Design of experimentation (DOE) is a planning process that serves to meet specific objectives. Proper planning of an experiment is very important in achieving the objectives clearly and efficiently with the right type of data and appropriate sample size. Many factors affect the performance of the heat transfer process in a heat exchanger. This chapter presents the discussion about design of experimentation in detail and generating design data for evolving experimental data-based models for various dependent/response variables of heat transfer performance in a heat exchanger through experimentation. Experiments are performed in almost any field of research and are used to study the performance of processes and systems. The process is a combination of machines, methods, people and other resources that transform some input into an output that has one or more observable responses. Some of the process variables are controllable, whereas other variables are uncontrollable, although they may be controllable for the purpose of a test. The experimental approach includes the following:

1. Determining which variables are most influential in the response.
2. Determining where to set the influential controllable variables so that the response is almost always near the desired optimal value.
3. The variability in the response is small.
4. The effect of uncontrollable variables is minimized.

The goal of any experimental activity is to get the maximum information about a system with the minimum number of well-designed experiments. An experimental program recognizes the major "factors" that affect the outcome of the experiment. The factors may be identified by looking at all the quantities that may affect the outcome of the experiment. The most important among these may be identified using a few exploratory experiments, or from past experience, or based on some underlying theory or hypothesis. One has to choose the number of levels for each of the factors. The data can be recorded for the values of the factors by performing the experiments and maintaining the levels at these values.

Adoption of basic laws of mechanics could be applied for correlation of various dependent and independent parameters of the heat transfer process in a heat exchanger in a theoretical approach. A theoretical approach can be adopted in a case if known logic can be applied correlating the various dependent and independent parameters of the system. Though qualitatively, the relationship between dependent and independent variables could be known based on available literature data. The generalized quantitative relationship may not be known sometimes due to complexity of the phenomenon.

The experimental optimization of the heat transfer process in a heat exchanger by using nanofluid is complex. Hence, the formulation of a quantitative relation based on logic is not possible. And given that formulating a theoretical – i.e., logic-based – model is not possible, the only alternative is to formulate an experimental data-based model, using an experimental approach. Such a model is proposed in this investigation. Hilbert Schenck Jr. has suggested a methodology of experimentation approach to formulating experimental data-based models for predicting behavior of such complex phenomena as heat transfer performance in a heat exchanger.

The design of the experiment involves the following steps:

1. First is identifying the independent and dependent variables which affect the phenomenon, based on the known qualitative physical characteristics of the phenomenon. The experimentation becomes time consuming, tedious and costly if it involves a large number of independent variables. With the help of dimensional analysis, one can reduce the number of variables, hence reducing the number of dimensional equations in the form of mathematical models.
2. Test planning consists of deciding the test envelope, the test sequence and the plan of experimentation for the set of deduced dimensional equations.

It is necessary to evolve the physical design of the experimental setup to set the test points, adjust the test sequence and execute the proposed experimental plan. Note the responses and provision for necessary instrumentation for deducing the relation of dependent pi terms of the dimensional equation in terms of independent pi terms. Experimental setup is designed in such a way that it can accommodate the range of independent and dependent variables within the proposed test envelope of the experimental plan. After recording the responses and obtained dimensional relations of dependent pi terms of dimensional equations, the exact mathematical model can be formed within the specified test envelope.

3.3 EXPERIMENTAL SETUP AND PROCEDURE

The two experimental sets are designed for the purpose of carrying out the experimentation to investigate and validate the phenomenon.

a. Two-wire method
b. Radiator as a heat exchanger

3.4 TWO-WIRE METHOD: EXPERIMENTAL PROCEDURE

The hot wire method is a standard technique to determine the thermal conductivity of nanofluid based on the measurement of the temperature rise in a definite distance from a linear hot wire immersed in the nanofluid. The hot

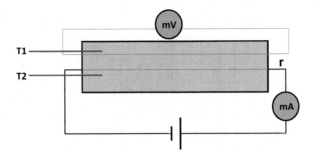

Figure 3.1 Schematic Sketch of Two-Wire Method–Based Setup to Determine Thermal Conductivity of Nanofluid.

wire probe method is based on the principle of transient hot wire method. If the output of the hot wire immersed in the nanofluid is constant and uniform, then thermal conductivity can be directly measured using change in the temperature over a known time interval.

1. The wire, which is immersed in the investigated sample, is heated by passing a constant electrical current through it, and its temperature change (ΔT) is measured as a function of time (t).
2. The setup consists of a chemically inert, cylindrical hallow tube of Teflon with a special arrangement of wire and thermocouple. The schematic representation of the setup is shown in Figure 3.1.
3. Silver wire of 200 mm and 2 mm diameter is used as the metal wire.
4. The required quantity of fluid volume to cover the silver wire is filled in the tube before the start of the experiment.
5. In this experimental setup, the current can be varied between 50 and 110 mA. The milli-Voltmeter and milli-Ammeter are used simply to display the value of voltage and current, respectively.
6. The experimentation is performed at room temperature for fixed values of controls to avoid uninvited fluctuations in the curves (ΔT versus time).
7. Thermal conductivity of nanofluids was estimated by using the relation,

$$k = (q/4\pi K)$$

where K is the slope between ΔT and time, k is thermal conductivity (W/mK) and q is the power generated by heating the wire per unit length (W/m).

3.5 RADIATOR AS A HEAT EXCHANGER: EXPERIMENTAL PROCEDURE

The second experimental setup is shown in Figure 3.2. Whereas with the two-wire method, the experimental setup fluid under investigation is in

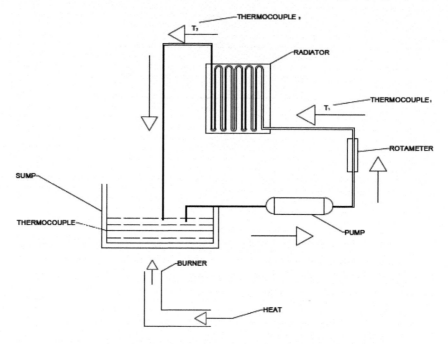

Figure 3.2 Schematic Diagram of Radiator as a Heat Exchanger.

static position, in the radiator method, the fluid under investigation is in motion.

Here, the radiator as a heat exchanger experimental setup uses water, engine coolant and Al_2O_3 – water nanofluid one by one as fluid under investigation. Mass flow rate readings are taken in the range of 0.5 to 5 lit/min (with the step of 0.5). The brief procedure is as follows:

1. Readings of temperature difference ΔT are taken for water, engine coolant and nanofluid.
2. Readings for Al_2O_3-water nanofluid at 0.5, 1.0, 1.5, 2.0 and 2.5 wt % concentration, 60 min. probe sonication (for size) and cubic shape nanoparticles are tabulated.
3. Readings for Al_2O_3-water nanofluid at 2.5 wt % concentration, cubic shape nanoparticles at 15, 30, 45 and 60 min. probe sonication (size) are tabulated.
4. Readings for Al_2O_3-water nanofluid at 2.5 wt % concentration, 60 min probe sonication (size) and cubic, spherical and cylindrical (rod) shape nanoparticles are tabulated.

3.6 DESIGN OF INSTRUMENTATION FOR EXPERIMENTAL SETUP

An instrumentation system was designed to measure fluid discharge, temperature of inlet and outlet fluid, and water quantity. Figure 3.3 shows the schematic diagram of the designed instrumentation for the experimental setup.

3.7 COMPONENTS OF INSTRUMENTATION SYSTEMS

1. Fluid Pump   2. Thermocouples   3. Radiator Fan
4. Rotameter   5. Switch Board   6. Heater
7. Radiator   8. Water Tank

3.8 IDENTIFICATION OF VARIABLES IN PHENOMENA

The variables affecting the heat transfer performance in a heat exchanger under consideration are concentration, shape, size, temperature, heat flow, heat transfer coefficient, density and mass flow rate.

Figure 3.3 Pictorial View of Radiator as a Heat Exchanger Experimental Setup.

Table 3.1 Various Dependent and Independent Variables with Symbols, Units, Dimensions, Nomenclature and Nature for Two-Wire Method

S. N.	Variables	Symbol	Unit	Nature (Dependent/ Independent)
1	Thermal conductivity based on concentration	K_ϕ	W/m°K	Dependent
2	Thermal conductivity based on size	K_t	W/m°K	Dependent
3	Thermal conductivity based on shape	K_s	W/m°K	Dependent
4	Density of nanoparticles	ρ	kg/m³	Independent
5	Temperature of nanofluid	T	°K	Independent
6	Prob sonication time	t	Sec	Independent
7	Size of nanoparticles	S	nm	Independent
8	Concentration of nanofluid	ϕ	%	Independent

Table 3.2 Various Dependent and Independent Variables with Symbols, Units, Dimensions, Nomenclature and Nature for Experimental Setup (Radiator as Heat Exchanger)

S. N.	Variables	Symbol	Unit	Nature (Dependent/ Independent)
1	Total heat flow	Q	W	Dependent
2	Coefficient of heat transfer	h	W/m² K	Dependent
3	Temperature difference	ΔT	°K	Dependent
4	Density of nanoparticles	ρ	kg/m³	Independent
5	Mass flow rate	m_f	kg/s	Independent
6	Probe sonication time	t	Sec	Independent
7	Size of nanoparticles	S	nm	Independent
8	Concentration of nanofluid	ϕ	%	Independent

3.9 MATHEMATICAL RELATIONSHIP FOR HEAT TRANSFER PHENOMENA

The measurement of inlet and outlet temperature of fluid in the radiator, by thermocouple measurement of surface and ambient temperature, by thermocouple measurement of mass flow rate of fluid under investigation by Rotameter

Measurement of temperature difference (ΔT), heat flow (Q) and heat transfer coefficient (h)

Total heat flow,

$$Q = m_f \times C_p \times (T_i - T_o) \text{ Watt}$$

where m_f = mass of liquid flow in Kg/s, Cp = specific heat of fluid kJ/kg °K, T_i = inlet temperature of fluid, °K, and T_o = outlet temperature of fluid in °K
   Also, $Q = h \times A \times (T_s - T_a)$
   Therefore, the heat transfer coefficient is H = Q/A x $(T_s - T_a)$, W/m² °K,
   where A = area of flow, i.e., A = N × π × d × l, N = number of tubes, d = diameter of tube in m, l = length of tube in m, T_s = surface temperature of tube in °K and T_a = atmospheric temperature in °K

3.10 FORMATION OF PI TERMS FOR DEPENDENT AND INDEPENDENT

Three independent pi terms (π_1, π_2, π_3) and three dependent pi terms (π_{D1}, π_{D2}, π_{D3}) have been identified in the design of experimentation and are available for the model formulation. Each dependent π term is assumed to be function of the available independent π terms.

3.10.1 For Two-Wire Method

$$K_\varphi = f\ (\rho, T, \varphi)$$

$$\pi_{D1} = f_1(\pi_1, \pi_2, \pi_3) \qquad \pi_{D2} = f_1(\pi_1, \pi_2, \pi_3) \qquad \pi_{D3} = f_1(\pi_1, \pi_2, \pi_3)$$

Where f is the function of a probable exact mathematical form. In this phenomenon, the empirical relationship between dependent and independent terms are assumed to be exponential. For example, the model representing the relationship between the dependent π term and independent π term is assumed to be as under.

$$\left(K_\varphi\right) = f_1\left\{(\rho)(T)(\varphi)\right\} \tag{3.1}$$

$$\left(K_t\right) = f_1\left\{(\rho)(T)(t)\right\} \tag{3.2}$$

$$\left(K_s\right) = f_1\left\{(\rho)(T)(s)\right\} \tag{3.3}$$

Model of dependent π term for thermal conductivity based on concentration, shape and size is as follows:

$$\pi_{D1} = k_1 \times (\pi_1)^{a1} \times (\pi_2)^{b1} \times (\pi_3)^{c1} \tag{3.4}$$

$$\pi_{D2} = k_2 \times (\pi_1)^{a2} \times (\pi_2)^{b2} \times (\pi_3)^{c2} \tag{3.5}$$

$$\pi_{D3} = k_3 \times (\pi_1)^{a3} \times (\pi_2)^{b3} \times (\pi_3)^{c3} \tag{3.6}$$

3.10.2 For Radiator as a Heat Exchanger

$$\Delta T = f(\varphi, \rho, t, S, m_f)$$

$$\pi_{D1} = f_1(\pi_1, \pi_2, \pi_3, \pi_4, \pi_5) \qquad \pi_{D2} = f_1(\pi_1, \pi_2, \pi_3, \pi_4, \pi_5)$$
$$\pi_{D3} = f_1(\pi_1, \pi_2, \pi_3, \pi_4, \pi_5)$$

For example, the model representing the relationship between the dependent and independent π terms assumed to be as follows:

$$(\Delta T) = f_1\{(\varphi)(\rho)(t)(S)(m_f)\} \tag{3.7}$$

$$(Q) = f_2\{(\varphi)(\rho)(t)(S)(m_f)\} \tag{3.8}$$

$$(h) = f_3\{(\varphi)(\rho)(t)(S)(m_f)\} \tag{3.9}$$

The models of the dependent pi terms for temperature difference, heat flow and heat transfer coefficient are

$$\pi_{D1} = k_1 \times (\pi_1)^{a1} \times (\pi_2)^{b1} \times (\pi_3)^{c1} \times (\pi_4)^{d1} \times (\pi_5)^{e1} \tag{3.10}$$

$$\pi_{D2} = k_2 \times (\pi_1)^{a2} \times (\pi_2)^{b2} \times (\pi_3)^{c2} \times (\pi_4)^{d2} \times (\pi_5)^{e2} \tag{3.11}$$

$$\pi_{D3} = k_3 \times (\pi_1)^{a3} \times (\pi_2)^{b3} \times (\pi_3)^{c3} \times (\pi_4)^{d3} \times (\pi_5)^{e3} \tag{3.12}$$

3.11 REDUCTION OF VARIABLES BY DIMENSIONAL ANALYSIS

Deducing the dimensional equation for a phenomenon reduces the number of independent variables in the experiments. The exact mathematical form of this dimensional equation is the targeted model. This is achieved by applying Buckingham's π theorem when n (no. of variables) is large. In the present research, there are only five independent variables and three dependent variables; hence, the process of reduction of variables is not necessary. Instead, empirical relations are established to find out indices of independent variables.

3.12 PLAN FOR EXPERIMENTATION

Deciding the test envelope, test points, test sequence and experimental plan for deduced set of dimensional equations is called as test planning or test plan of experiment.

3.12.1 Determination of Test Envelope

Deciding the end points or limits comprises the test envelope, and it is the obvious and best way to select or fix the test points. All such test points are covered in the test envelope. So, it is obvious to ascertain the complete range over which the entire experimentation is carried out. Table 3.3 shows the ranges for the various pi terms.

3.12.2 Determination of Test Points

The spacing of the test points within the test envelope is selected not for having a "symmetrical" or "pleasing" curve but to have every part of the experimental curve map the same level of precision as every other part. Spacing the independent variables in a predetermined manner and most effectively will constitute an efficient and compact experimental plan. Ideally, every part of the curve should be represented by test points but in an actual situation, constraints are imposed by the experimental setup. Hence, the conceptual proper spacing of test points is now replaced by permissible spacing of test points. The decided test points for heat transfer performance in heat exchanger by using nanofluids are shown in Table 3.3.

3.12.3 Determination of Test Sequence

The test sequence is chosen by reversible or irreversible nature of experiments. Basically, all the tests are irreversible in a sense; no apparatus can be brought to the original state of configuration after use. If the changes imposed by testing are below the level of detection, such tests can be considered reversible. In a classical or sequential plan, the variables are varied from one extremity to the other in a sequential manner; in the random plan, they are varied in random fashion. The classical or sequential plan is essentially selected for irreversible experiments or where there is no scope for randomization. The majority of engineering experiments use a partial randomized plan, and hence that is used for testing the phenomenon.

3.12.4 Determination of Test Plan for Experiment

The classical plan of experimentation is used in the present research. In this plan, all but one variable is maintained at their fixed levels, and only one under consideration is varied over its widest range decided by test points taken one

Table 3.3 Test Envelope, Test Points for Two-Wire Method

(a) For Concentration Model:

Test Point			Plan of Experimentation		
$\pi_1 = \rho$	$\pi_2 = T$	$\pi_3 = \Phi$	$\pi_1 = \rho$	$\pi_2 = T$	$\pi_3 = \Phi$
1018	303	0.0001	Fluid (Const)	Temp (Const)	Concentration (Vary 5)
1038	308	0.5	Fluid (Const)	Temp (Vary 11)	Concentration (Const)
1106	313	1	Fluid (Vary 6))	Temp (Const)	Concentration (Const)
1120	318	1.5	5	11	6
1123	323	2		Total readings: 5 × 11 × 6 = 330	
	328	2.5			
	333				
	338		Test Envelope		
	343		$\pi_1 = \rho$	$\pi_2 = T$	$\pi_3 = \Phi$
	348		1018 to	303 to	0.0001 to
	353		1123	353	2.5

(b) For Particle Size Model:

Test Point			Plan of Experimentation		
$\pi_1 = \rho$	$\pi_2 = T$	$\pi_3 = t$	$\pi_1 = \rho$	$\pi_2 = T$	$\pi_3 = t$
1018	303	15	Fluid (Const)	Temp (Const)	Particle Size (Vary 4)
1038	308	30	Fluid (Const)	Temp (Vary 11)	Particle Size (Const)
1106	313	45	Fluid (Vary 6))	Temp (Const)	Particle Size (Const)
1120	318	60	5	11	4
1123	323			Total readings: 5 × 11 × 4 = 220	
	328				
	333		Test Envelope:		
	338		$\pi_1 = \rho$	$\pi_2 = T$	$\pi_3 = t$
	343		1018 to	303 to	15 to
	348		1123	353	60
	353				

(Continued)

(c) For Particle Shape Model:

Test Point			Plan of Experimentation		
$\pi_1 = \rho$	$\pi_2 = T$	$\pi_3 = S$	$\pi_1 = \rho$	$\pi_2 = T$	$\pi_3 = S$
1018	303	14.23	Fluid (Const)	Temp (Const)	Particle Size (Vary 4)
1038	313	18.57	Fluid (Const)	Temp (Vary 11)	Particle Size (Const)
1106	323	64.28	Fluid (Vary 6)	Temp (Const)	Particle Size (Const)
1120	333		5	6	3
1123	343		Total readings: 5 × 6 × 3 = 90		
	353		*Test Envelope:*		
			$\pi_1 = \rho$	$\pi_2 = T$	$\pi_3 = S$
			1018 to 1123	303 to 353	14.23 to 64.28

at a time. In this study, only one independent pi term is varied over its planned test points' range and all other independent pi terms are kept constant.

The readings of temperature difference for water, engine coolant and nano-fluid versus discharge will be continuously recorded each time using the planned instrumentation system for necessary analysis. During each experiment the earlier independent pi term is set at its fixed value and one of the other independent pi terms is varied over its range. Dependent variables are required to be evaluated such as thermal conductivity at concentration, shape and size.

3.13 EXPERIMENTAL OBSERVATIONS

3.13.1 For Two-Wire Method

3.13.2 For Radiator as a Heat Exchanger

3.14 SAMPLE SELECTION

3.14.1 Sampling

The sampling method refers to the rules and procedures by which some elements of the population are included in the sample. In statistics, population represents an aggregation of objects, animate or inanimate, under study.

The population may be finite or infinite. A part or small portion selected from population is called "sample", and the process of such selection is called "sampling". Sampling is resorted to when it is not possible to enumerate the whole population or when it is too costly to enumerate in terms of money and time.

3.14.2 Sample Design

A sample design is a definite plan for obtaining a sample from a given population. It refers to the technique or procedure to be adopted in selecting items for the sample and the size of the sample. Sample design is determined before the data are collected. There are many sample designs and that design should be selected which is reliable and appropriate for particular research study.

3.14.3 Types of Sample Designs

There are different types of sampling designs and they are based on two factors: the representation basis and the element selection technique. On a representation basis, they are classified as probability sampling and non-probability sampling. Probability sampling is based on the concept of random selection, whereas non-probability sampling is non-random sampling. On an element selection basis, the sampling may be either unrestricted or restricted. When a sample element is drawn from a large population, then the sample so drawn is known as "unrestricted sample", whereas all other forms of sampling are covered under the term "restricted sampling". Thus, the sample designs are basically categorized in to two types – non-probability and probability sampling.

Representation Basis		
Element Basis	Probability Sampling (1)	Non-Probability Sampling (2)
(1) Unrestricted sampling	Simple random sampling	Haphazard sampling or convenience sampling
(2) Restricted sampling	Complex random sampling	Purposive sampling or judgment sampling

3.14.4 Sample Size for Experiments

Selection of proper sample size is an important aspect while conducting an experimental study. The recorded readings of the experimental study are mentioned in Tables 3.4 to 3.6. It is imperative to estimate the sample size required for an assumed level of confidence. In case the size of sample has been assumed, the analyst has to estimate the confidence limits that they could attach for the given level of confidence.

An important question in sampling is what should be the sample size. Assuming each repetition of the experiment under the same conditions

Table 3.4 Test Envelope, Test Points for Experimental Setup of Radiator as a Heat Exchanger

Test Point

ϕ	ρ	t	S	m_f
%	kg/m^3	Sec	m	(kg/s)
0.5	1061	900	5.19E-08	0.008333
1	1073	1800	2.29E-08	0.016667
1.5	1085	2700	4.31E-08	0.025
2	1097	3600		0.033333
2.5	1109	4500		0.041667
				0.05
				0.058333
				0.066667
				0.075
				0.08333

Test Envelope

ϕ	ρ	t	S	m_f
0.5 to 2.5	1061 to 1109	900 to 4500	2.29E-08 to 5.19E-08	0.008333 to 0.083333

Plan of Experimentation

ϕ	ρ	t	S	m_f
Concentration	Density	Size	Shape	Flow rate
Vary	Const	Const	Const	Const
Const	Vary	Const	Const	Const
Const	Const	Vary	Const	Const
Const	Const	Const	Vary	Const
Const	Const	Const	Const	Vary

Table 3.5 Thermal Conductivity Based on Concentration, $K\varphi$

Consolidated Concentration Sheet

S.N.	Fluid	$\pi_1 = \rho$ Density	$\pi_2 = T$ T (K)	$\pi_3 = \Phi$ Concentration	$Z = K\varphi$ K
		Kg/m^3	K	%	$W/m°K$
1	Al_2O_3	1038	303	0.0001	0.5612
2	Al_2O_3	1038	303	0.5	0.772
3	Al_2O_3	1038	303	1	0.85

(Continued)

Table 3.5 (Continued)

Consolidated Concentration Sheet

S.N.	Fluid	$\pi_1 = \rho$ Density Kg/m^3	$\pi_2 = T$ T (K) K	$\pi_3 = \Phi$ Concentration %	$Z = K\varphi$ K $W/m°K$
4	Al_2O_3	1038	303	1.5	0.924
5	Al_2O_3	1038	303	2	1.049
6	Al_2O_3	1038	303	2.5	1.16
7	Al_2O_3	1038	308	0.0001	0.5637
8	Al_2O_3	1038	308	0.5	0.798
9	Al_2O_3	1038	308	1	0.876
10	Al_2O_3	1038	308	1.5	0.963
11	Al_2O_3	1038	308	2	1.08
12	Al_2O_3	1038	308	2.5	1.188
13	Al_2O_3	1038	313	0.0001	0.5662
14	Al_2O_3	1038	313	0.5	0.82
15	Al_2O_3	1038	313	1	0.906
16	Al_2O_3	1038	313	1.5	0.993
17	Al_2O_3	1038	313	2	1.114

Table 3.6 Observations Related to ΔT, Q and h

S.N.	Concent-ration	Density (Kg/m^3)	Probe sonication time in sec	Size (m)	DISCHARGE (Liter/Sec)	DIFFERENCE $(T1 - T2)$	$Q =$ $mf*Cp*\Delta T$	Heat Transfer Rate $h = Q/A$ $(Ts-Ta)$
	π_1	π_2	π_3	π_4	π_5	π_{D1}	π_{D2}	π_{D3}
0	$\pi_1 = \varphi$	$\pi_2 = \rho$	$\pi_3 = T$	$\pi_4 = S$	$\pi_5 = m_f$	$\Delta T =$ $(T_i - T_o)$	Q	h
0	%	kg/m^3	Sec	m	kg/s	°K	W	$W/m^2 °K$
1	0.5	1061	3600	5.19E-08	0.008333	12.5	0.427083	0.049382
2	0.5	1061	3600	5.19E-08	0.008333	12.4	0.423667	0.050749
3	0.5	1061	3600	5.19E-08	0.008333	10.5	0.35875	0.048961
4	0.5	1061	3600	5.19E-08	0.008333	11.5	0.392917	0.055207
5	0.5	1061	3600	5.19E-08	0.008333	10.7	0.365583	0.057424
6	0.5	1061	3600	5.19E-08	0.016667	12.7	0.867833	0.097631
7	0.5	1061	3600	5.19E-08	0.016667	13.5	0.9225	0.110501
8	0.5	1061	3600	5.19E-08	0.016667	13.1	0.895167	0.112064
9	0.5	1061	3600	5.19E-08	0.016667	11.9	0.813167	0.109187
10	0.5	1061	3600	5.19E-08	0.016667	10.6	0.724333	0.106257
11	0.5	1061	3600	5.19E-08	0.025	12.5	1.28125	0.141746
12	0.5	1061	3600	5.19E-08	0.025	12.7	1.30175	0.148453
13	0.5	1061	3600	5.19E-08	0.025	12.3	1.26075	0.150477
14	0.5	1061	3600	5.19E-08	0.025	12.3	1.26075	0.15724
15	0.5	1061	3600	5.19E-08	0.025	11	1.1275	0.155792
16	0.5	1061	3600	5.19E-08	0.033333	10.3	1.407667	0.166816

Table 3.7 Sample Size Calculation for 10 Samples

S.N.	$Xi = \Delta T =$ $(T_s - T_a)$ (Expm)	μ	$d = (X_i - \mu)$	$(X_i - \mu)^2$
1	28.8	25.74	3.06	9.3636
2	27.8	25.74	2.06	4.2436
3	24.4	25.74	1.34	1.7956
4	23.7	25.74	2.04	4.1616
5	21.2	25.74	4.54	20.6116
6	29.6	25.74	3.86	14.8996
7	27.8	25.74	2.06	4.2436
8	26.6	25.74	0.86	0.7396
9	24.8	25.74	0.94	0.8836
10	22.7	25.74	3.04	9.2416
				σ^2
Average	**25.74**		**Sum**	70.184
				Σ
	d	Avg	**2.38**	**8.377589**
	d	Max	**4.54**	
	d	Min	**0.86**	

For d = davg, n = 47.59 ≈ 48
For d = dmax, n = 13.0809 ≈ 14
For d = dmin, n = 364.54 ≈ 365

given one value; the question is how many times the experiment has to be repeated. To clarify this, one should have the information of (1) the nature of population, (2) type of sampling, (3) the nature of study, (4) the standard of accuracy or precision desired, (5) the cost of conducting each experiment, (6) the cost of imprecise results, (7) the variability of process, and (8) the amount of confidence required for the result of experiment.

As a general rule the sample must be of optimum size. The sample size "n" for the confidence level $(1 - \alpha)$ with the interval on either side of the mean as ±d, is calculated using the relation given below if the variance of the population σ^2 is known.

$$n = \left[\frac{(Z_{\alpha/2})(\sigma)}{d} \right]^2$$

where d denotes the error range that decreases with the increase in sample size. Therefore, for a given error size, one can estimate the sample size required for the experiment. The performer of the experiment has to make a reasonable assumption for a value of d. $Z\alpha/2$ is the standardized normal statistics for the probability we seek.

In many cases, the exact value of σ may not be known. Suppose we have an idea of the highest and lowest possible result of the experiment (feasible range); we can assume the range approximately equal to 4σ to estimate σ.

If the variance is not known, a pilot experiment may be conducted and based on the results, an estimate of variance (s^2) can be computed. The t-distribution can be used and sample size can be estimated assuming a reasonable value for d.

$$n = \left[\frac{(t_{\alpha/2})(S)}{d} \right]^2,$$

where t is the tabulated t-value for the desired confidence level and degrees of freedom of initial sample, and d is the half-width of desired confidence interval assumed.

Therefore, the sample size for the present investigation is calculated as follows:

$$n = \left[\frac{(Z_{\alpha/2})(\sigma)}{d} \right]^2.$$

We take a 95% confidence level, From the normal table for 95% probability,

$$Z = 1.96.$$

Assuming the estimate should lie within the interval $n = \mu \pm \left(\dfrac{\sigma}{5} \right)$ with probability of 0.95 where μ is population mean,

$$n = \left[\frac{(1.96)(\sigma)}{\dfrac{\sigma}{5}} \right]^2$$

$$n = 96.04 \approx 97.$$

The actual sample size in this investigation is increased by 10% so that the error in the phenomenon can decrease with an increase in sample size.

The mean μ and variance σ^2 can be calculated as follows:

$$\mu = \frac{1}{n} \sum_{i=1}^{n} X_i$$

$$\sigma^2 = \frac{1}{n} \sum_{i=1}^{n} (X_i - \mu)^2$$

Thus, standard deviation is given as:

$$\sigma = \sqrt{\frac{1}{n}\sum_{i=1}^{n}(X_i - \mu)^2}$$

For 10 samples, $\mu = 25.74$, $\sigma = 8.3777$ and $d_{(avg)} = 2.38$

$$n = \left[\frac{(1.96)(8.3777)}{2.38}\right]^2$$

$n = 47.598 \approx 48$

For d = dmax, n = 13.0809 ≈ 14
For d = dmin, n = 364.54 ≈ 365

Thus, the sample size of 550 considered in this experimentation satisfactorily matches the criteria of deciding the sample size.

Chapter 4

Mathematical Models

4.1 INTRODUCTION

Engineers are interested in designing devices and processes and systems. That is, beyond observing how the world works, engineers are interested in creating artifacts that have not yet come to life. Thus, engineers must be able to describe and analyze objects and devices in order to predict their behavior and to see if that behavior is what the engineers want. In short, engineers need to model devices and processes if they are going to design them. Mathematical modelling is a *principled* activity that has both principles behind it and methods that can be successfully applied. The principles are overarching or *meta*-principles, phrased as questions about the intentions and purposes of mathematical modelling. Although the scientific method and engineering design have much in common, there are differences in motivation and approach that are worth mentioning. In the practices of both science and engineering design, models are often applied to predict what will happen in a future situation. In engineering design, however, the predictions are used in ways that have far different consequences than simply anticipating the outcome of an experiment.

4.2 MODEL CLASSIFICATION

Mathematical models can be classified in several ways:

1. *Linear model*: If all the operators in a mathematical model exhibit linearity, the resulting mathematical model is defined as linear. Also, if the objective functions and constraints are represented entirely by linear equations, then the model is regarded as a linear model.
2. *Nonlinear model*: If one or more of the objective functions or constraints are represented with a nonlinear equation, then the model is known as a nonlinear model. It is often associated with phenomena such as chaos and irreversibility. Although there are exceptions,

DOI: 10.1201/9781003432111-4

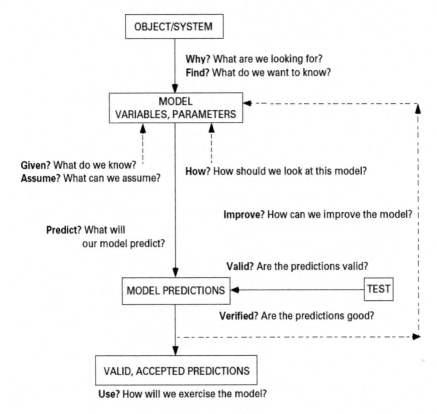

Figure 4.1 Principles of Mathematical Modelling.

nonlinear systems and models tend to be more difficult to study than linear ones. A common approach to nonlinear problems is linearization, but this can be problematic if one is trying to study aspects such as irreversibility, which is strongly tied to nonlinearity.

3. *Static model*: A *static* (or steady-state) model calculates the system in equilibrium, and thus is time-invariant.

4. *Dynamic model*: A *dynamic* model accounts for time-dependent changes in the state of the system. Dynamic models typically are represented by differential equations.

5. *Explicit model*: If all the input parameters of the overall model are known, and the output parameters can be calculated by a finite series of computations the model is said to be explicit model.

6. *Implicit model*: If the output parameters are known, and the corresponding inputs must be solved for by an iterative procedure, such as Newton's method (if the model is linear) or Broyden's method (if nonlinear), the model is said to be implicit.

7. *Discrete model*: A discrete model treats objects as discrete, such as the particles in a molecular model or the states in a statistical model.

8. *Continuous model*: A continuous model represents the objects in a continuous manner, such as the velocity of fluid in pipe flows, temperatures and stresses in a solid, and an electric field that applies continuously over the entire model due to a point charge.

9. *Deterministic model*: A deterministic model is one in which every set of variable is uniquely determined by parameters in the model and by sets of previous states of these variables. A deterministic model always performs the same way for a given set of initial conditions.

10. *Probabilistic (stochastic) model*: A stochastic model is usually called a "statistical model" where randomness is present, and variable states are not described by unique values, but rather by probability distributions

11. *Deductive, inductive, or floating models*: A deductive model is a logical structure based on a theory. An inductive model arises from empirical findings and generalization from them. The floating model rests on neither theory nor observation, but is merely the invocation of expected structure. Application of mathematics in social sciences outside of economics has been criticized for unfounded models.

4.3 FORMULATION OF EXPERIMENTAL DATA-BASED MODELS (TWO-WIRE METHOD)

4.3.1 Model of Dependent Pi Term for π_{D1} (K_Φ) (Concentration)

$$\pi_{D1} = f\left(\pi_1, \pi_2, \pi_3\right)$$

Approximate generalized experimental models for predicting thermal conductivity based on concertation into wire method is given by

$$\pi_{D1} = K_1 \times (\pi_1)^{a1} \times (\pi_2)^{b1} \times (\pi_3)^{c1}. \tag{4.1}$$

The values of exponential a_1, b_1, c_1 are established, considering the exponential relationship between the dependent pi term K_Φ and independent π terms π_1, π_2, π_3, independently taken one at a time on the basis of data collected through classical experimentation. Thus, corresponding to the three independent pi terms one has to formulate three pi terms from the set of observed data for thermal conductivity based on concentration, K_Φ. From these, models' values of dependent pi terms are computed. There are four unknown terms in Equation (4.1) – the curve-fitting constant K_1 and indices

a_1, b_1, c_1. To get the values of these unknown terms, we need at minimum four sets of values of (π_1, π_2, π_3).

As per the experimental plan in design of experimentation, 540 sets of these values are recorded. If any arbitrary 20 sets from table are selected and the values of unknown K_1 and indices a_1, b_1, c_1 are computed, then it may not result in one best unique solution representing a best fit unique curve for the remaining sets of values. To be very specific, one can find ^nC_r combinations of r that are taken together out of the available n sets of the values. The value ^nC_r in this case will be $^{540}C_3$.

Solving these many sets and finding their solutions will be a herculean task. Hence, this problem will be solved using the curve fitting technique. To follow this method it is necessary to have the equations in the form as follows:

$$Z = a + bX + cY + dZ \tag{4.2}$$

Equation (4.1) can be brought in the form of Equation (4.2) by taking log on both sides of the equation for π_{D1}, to get four unknown terms in the equations,

$$\text{Log } \pi_{D1} = \log k_1 + a_1 \log \pi_1 + b_1 \log \pi_2 + c_1 \log \pi_3 \tag{4.3}$$

Let, $Z_1 = \log \pi_{D1}$, $K_1 = \log k_1$, $A = \log \pi_1$, $B = \log \pi_2$, $C = \log \pi_3$

Putting the values in Equation (2), the same can be written as

$$Z_1 = K_1 + a_1 A + b_1 B + c_1 C \tag{4.4}$$

Equation (4.4) is a regression equation of Z on A, B and C in an n-dimensional coordinate system. This represents a regression hyper plane. To determine the regression hyper plane, determine a_1, b_1, c_1 in Equation (4.4) so that,

$$\sum Z_1 = nK_1 + a_1^* \sum A + b_1^* \sum B + c_1^* \sum C$$

$$\sum Z_1^* A = K_1^* \sum A + a_1^* \sum A^* A + b_1^* \sum B^* A + c_1^* \sum C^* A$$

$$\sum Z_1^* B = K_1^* \sum B + a_1^* \sum A^* B + b_1^* \sum B^* B + c_1^* \sum C^* B$$

$$\sum Z_1^* C = K_1^* \sum C + a_1^* \sum A^* C + b_1^* \sum B^* C + c_1^* \sum C^* C \tag{4.5}$$

In Equation (4.5), the values of the multipliers K_1, a_1, b_1 and c_1 are substituted to compute the values of the unknowns (viz. K_1, a_1, b_1 and c_1). The values of the terms on L.H.S. and the multipliers of K_1, a_1, b_1 and c_1 in the set of equations are calculated. After substituting these values in Equation (4.1), one will get a set of four equations, which are to be solved simultaneously to get the values of K_1, a_1, b_1 and c_1. Equation (4.5) can be verified in the matrix form and further values of K_1, a_1, b_1 and c_1 can be obtained by using matrix analysis.

$$X_1 = \text{inv}(W) \times P_1 \tag{4.6}$$

The matrix method of solving these equations using MATLAB® is as follows:

W = 4 × 4 matrix of the multipliers of K_1, a_1, b_1 and c_1,
P_1 = 4 × 1 matrix of the terms on L.H.S. and
X_1 = 4 × 1 matrix of solutions of values of K_1, a_1, b_1 and c_1.
Then, the matrix obtained is given by,

$$Z_1 \begin{bmatrix} 1 \\ A \\ B \\ C \end{bmatrix} = \begin{bmatrix} n & A & B & C \\ A & A^2 & BA & CA \\ B & AB & B^2 & CB \\ C & AC & BC & C^2 \end{bmatrix} \begin{bmatrix} K_1 \\ a_1 \\ b_1 \\ c_1 \end{bmatrix}$$

$$P_1 = W_1 \times X_1$$

−29.2769		330	1001.042	830.0715	−188.428	K_1
−88.8859	=	1001.042	3036.725	2517.988	−571.59	× a_1
−73.4151		830.0715	2517.988	2088.081	−473.967	b_1
55.00095		−188.428	−571.59	−473.967	900.3831	c_1

$$[P_1] = [W_1][X_1]$$

Using MATLAB, $X_1 = W_1/P_1$, after solving X_1 matrix with K_1 and indices a_1, b_1, c_1 values are as follows:

K_1	−1.8219
a_1	−0.7166
b_1	1.5642
c_1	0.483

But, K_1 is log value. So, to convert into normal value, taking the antilog of K_1:

Antilog $(-1.8219) = 1.51E{-}02$

Hence the model for dependent term π_{D1} is:

$$\pi_{D1} = K_1 \times (\pi_1)^{a1} \times (\pi_2)^{b1} \times (\pi_3)^{c1}$$

$$(K_\varphi) = K1\left\{(\rho)^{a1}\,(T)^{b1}\,(\varphi)^{c1}\right\}$$

$$(K_\varphi) = 1.51E - 02\left\{(\rho)^{-0.7166}\,(T)^{1.5642}\,(\varphi)^{0.483}\right\}. \tag{4.7}$$

4.3.2 Model of Dependent Pi Term for π_{D2} (Size, K_t)

$$P_2 = W_2 \times X_2$$

9.26489	220	667.3611	553.381	334.6517	K_2
28.05107 =	667.3611	2024.483	1678.659	1015.152	× a_2
23.54413	553.381	1678.659	1392.054	841.7722	b_2
16.73287	334.6517	1015.152	841.7722	520.3006	c_2

$$[P_2] = [W_2][X_2]$$

Using MATLAB, $X_2 = W_2/P_2$, after solving the X_2 matrix with K_2 and indices a_2, b_2, c_2 is as follows

K_2	−4.2329
a_2	−0.7615
b_2	2.476
c_2	0.2347

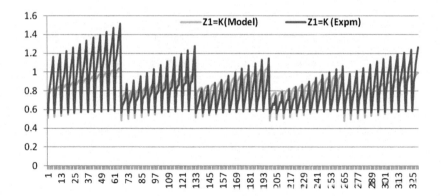

Figure 4.2 Comparison between Model and Experimental Data of Thermal Conductivity Based on Concentration ($K\phi$).

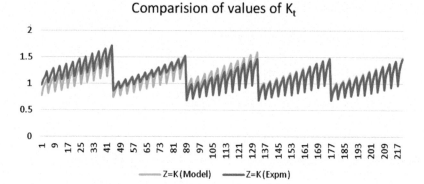

Figure 4.3 Comparison between Model and Experimental Data of Thermal Conductivity Based on Pro-Sonication Time, K_t.

But K_2 is log value. So, to convert into normal value, taking the antilog of K_2:

Antilog $(-4.2329) = 5.85\text{E} - 05$

Hence, the model for dependent term π_{D2}

$$\pi_{D2} = K_2 \times (\pi_1)^{a2} \times (\pi_2)^{b2} \times (\pi_3)^{c2}$$

$$K_t = K_2 \left\{ (\rho)^{a2} (T)^{b2} (t)^{c2} \right\}$$

$$K_t = 5.85\text{E} - 05 \left\{ (\rho)^{-0.7615} (T)^{2.476} (t)^{0.2347} \right\} \tag{4.8}$$

4.3.3 Model of Dependent Pi Term for π_{D3} (K_s) (Shape)

6.596922	165	501.759	415.0358	244.6352	K_3
19.95335	501.759	1525.892	1262.109	743.9379	a_3
16.71595	415.0358	1262.109	1044.04	615.3477	b_3
10.52121	244.6352	743.9379	615.3477	372.2715	c_3

$$[P_3] = [W_3][X_3]$$

Using MATLAB, $X_3 = W_3/P_3$, after solving X_3 matrix with K_3 and indices a_3, b_3, c_3 values are as follows

K_3	1.0219
a_3	−1.7556
b_3	1.6852
c_3	0.0794

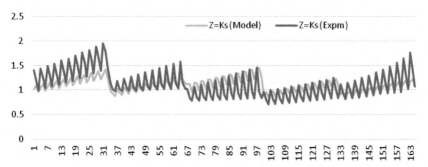

Figure 4.4 Comparison between Model and Experimental Data of Thermal Conductivity Based on Shape, K_s.

But K_1 is log value so to convert into normal value taking antilog of K_1 Antilog (1.0219) = 10.5172. Hence, the model for dependent term π_{D1} is:

$$\pi_{D3} = K_3 \times (\pi_1)^{a3} \times (\pi_2)^{b3} \times (\pi_3)^{c3}$$

$$(K_s) = 10.5172\left\{(\rho)^{-1.7556}\,(T)^{1.6852}\,(s)^{0.0794}\right\} \tag{4.9}$$

4.3.4 Model of Dependent Pi Term for π_{D1} (ΔT)

$$\pi_{D1} = f(\pi_1, \pi_2, \pi_3, \pi_4, \pi_5) \tag{4.10}$$

Approximate generalized experimental models for temperature difference for heat transfer performance in a heat exchanger have been established.

$$\pi_{D1} = k_1 \times (\pi_1)^{a1} \times (\pi_2)^{b1} \times (\pi_3)^{c1} \times (\pi_4)^{d1} \times (\pi_5)^{e1} \tag{4.11}$$

The values of exponential a_1, b_1, c_1, d_1 and e_1 are established, considering exponential relationship between the dependent pi term ΔT and independent π terms π_1, π_2, π_3, π_4, π_5 independently taken one at a time, on the basis of data collected through classical experimentation.

Thus, corresponding to the five independent pi terms, one has to formulate five pi terms from the set of observed data for temperature difference. From these models values of dependent πpi terms are computed.

There are six unknown terms in Equation (4.11). These are the curve fitting constant K_1 and indices a_1, b_1, c_1, d_1 and e_1. To get the values of these unknown we need minimum six sets of values of $(\pi_1, \pi_2, \pi_3, \pi_4, \pi_5)$.

As per the experimental plan in design of experimentation, we have 600 sets of these values. If any arbitrary 20 sets from table are selected and the values of unknown K_1 and indices a_1, b_1, c_1, d_1, e_1 are computed, then it may not result in one best unique solution representing a best fit unique curve for the remaining sets of values. To be very specific, one can find out ^nC_r combinations of r are taken together out of the available n sets of the values. The value ^nC_r in this case will be $^{600}C_5$,

Solving these many sets and finding their solutions will be a herculean task. Hence, the curve fitting technique will be used to solve this problem. To follow this method it is necessary to have the equations in the form as follows:

$$Z = a + bX + cY + dZ + \ldots\ldots \tag{4.12}$$

Equation (4.11) can be brought in the form of Equation (4.12) by taking log on both sides. Taking log on both sides of the equation for π_{D1} to get six unknown terms in the equations,

$$\text{Log } \pi_{D1} = \log k_1 + a_1 \log \pi_1 + b_1 \log \pi_2 + c_1 \log \pi_3 + d_1 \log \pi_4 \tag{4.13}$$
$$+ e_1 \log \pi_5$$

Let $Z_1 = \log \pi_{D1}$, $K_1 = \log k_1$, $A = \log \pi_1$, $B = \log \pi_2$,
$\quad C = \log \pi_3$, $D = \log \pi_4$, $E = \log \pi_5$

Putting the values in Equation (4.12), the same can be written as

$$Z_1 = K_1 + a_1 A + b_1 B + c_1 C + d_1 D + e_1 E \tag{4.14}$$

Equation (14) is a regression equation of Z on A, B, C, D and E, in an n-dimensional coordinate system. This represents a regression hyper plane. To determine the regression hyper plane, determine a_1, b_1, c_1, d_1, e_1, in Equation (4.13) so that,

$$\sum Z_1 = nK_1 + a_1^* \sum A + b_1^* \sum B + c_1^* \sum C + d_1^* \sum D + e_1^* \sum E$$

$$\sum Z_1^* A = K_1^* \sum A + a_1^* \sum A^* A + b_1^* \sum B^* A + c_1^* \sum C^* A + d_1^* \sum D^* A$$
$$+ e_1^* \sum E^* A$$

$$\sum Z_1^* B = K_1^* \sum B + a_1^* \sum A^* B + b_1^* \sum B^* B + c_1^* \sum C^* B + d_1^* \sum D^* B$$
$$+ e_1^* \sum E^* B$$

$$\sum Z_1^* C = K_1^* \sum C + a_1^* \sum A^* C + b_1^* \sum B^* C + c_1^* \sum C^* C + d_1^* \sum D^* C$$
$$+ e_1^* \sum E^* C$$

$$\sum Z_1^* D = K_1^* \sum D + a_1^* \sum A^* D + b_1^* \sum B^* D + c_1^* \sum C^* D + d_1^* \sum D^* D$$
$$+ e_1^* \sum E^* D$$

$$\sum Z_1^* E = K_1^* \sum E + a_1^* \sum A^* E + b_1^* \sum B^* E + c_1^* \sum C^* E + d_1^* \sum D^* E$$
$$+ e_1^* \sum E^* E \qquad (4.15)$$

The aforementioned equations can be verified in the matrix form and further values of
K_1, a_1, b_1, c_1, d_1 and e_1 can be obtained by using matrix analysis.

$$X_1 = \text{inv}(W) \times P_1 \qquad (4.16)$$

The matrix method of solving these equations using MATLAB is as follows:

$W = 6 \times 6$ matrix of the multipliers of $K_1, a_1, b_1, c_1, d_1, e_1$,
$P_1 = 6 \times 1$ matrix of the terms on L.H.S. and
$X_1 = 6 \times 1$ matrix of solutions of values of $K_1, a_1, b_1, c_1, d_1, e_1$

Then,
The matrix obtained is given by,

$$P_1 = W_1 \times X_1$$

442.1699	550	148.0836	1672.324	1909.41	−4028.58	−782.763 K_1
113.6503	148.0836	66.037	451.0388	508.1035	−1087.47	−210.753 a_1
1344.281 =	1672.324	451.0388	5084.872	5805.529	−12249.4	−2380.06 b_1
1539.028	1909.41	508.1035	5805.529	6648.779	−13987.7	−2717.48 c_1
−3238.13	−4028.58	−1087.47	−12249.4	−13987.7	29513.9	5733.498 d_1
−649.85	−782.763	−210.753	−2380.06	−2717.48	5733.498	1164.198 e_1

$$[P_1] = [W_1][X_1]$$

Using MATLAB, $X_1 = W_1/P_1$, after solving X_1 matrix with K_1 and indices a_1, b_1, c_1 are as follows:

K_1	38.1967
a_1	0.2028
b_1	−12.5534
c_1	0.1366
d_1	0.0457
e_1	−0.4097

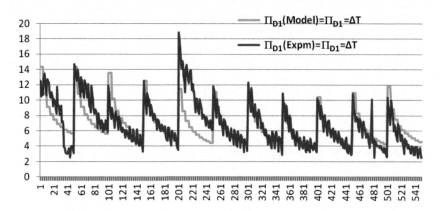

Figure 4.5 Comparison between Model and Experimental Data of Temperature Difference, ΔT.

But K_1 is log value. So, to convert into normal value, taking antilog of K_1 Antilog (38.1967) = 1.57E+38. Hence, the model for dependent term π_{D1} is:

$$\pi_{D1} = k_1 \times (\pi_1)^{a1} \times (\pi_2)^{b1} \times (\pi_3)^{c1} \times (\pi_4)^{d1} \times (\pi_5)^{e1}$$

$$(\Delta T) = f1\{(\varphi)(\rho)(T)(S)(mf)\}$$

$$(\Delta T) = 1.57\,E - 08\left\{(\varphi)^{0.2028}\,(\rho)^{-12.5534}\,(T)^{0.1366}\,(S)^{0.0457}\,(mf)^{-0.4097}\right\} \quad (4.17)$$

4.3.5 Model of Dependent Pi Term for π_{D2} (Q)

$$P_2 = W_2 \times X_2$$

−13.8192		550	148.0836	1672.324	1909.41	−4028.58	−782.763 K_2
−9.96637		148.0836	66.037	451.0388	508.1035	−1087.47	−210.753 a_2
−42.2188	=	1672.324	451.0388	5084.872	5805.529	−12249.4	−2380.06 b_2
−43.8144		1909.41	508.1035	5805.529	6648.779	−13987.7	−2717.48 c_2
101.9431		−4028.58	−1087.47	−12249.4	−13987.7	29513.9	5733.498 d_2
49.28196		−782.763	−210.753	−2380.06	−2717.48	5733.498	1164.198 e_2

$$[P_2] = [W_2][X_2]$$

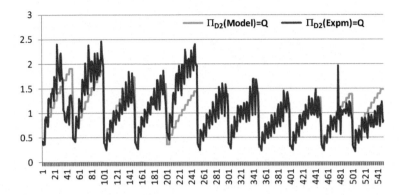

Figure 4.6 Comparison between Model and Experimental Data of Heat Flow, Q.

Using MATLAB, $X_2 = W_2/P_2$, after solving the X_2 matrix with K_2 and indices a_2, b_2, c_2, d_2 and e_2 is as follows:

K_2	38.7216
a_2	0.1698
b_2	−12.5273
c_2	0.1368
d_2	0.046
e_2	0.5903

But K_2 is log value. So, to convert into normal value taking antilog of K_2 Antilog (38.1967) = 5.27E+38, Hence, the model for dependent term π_{D2}:

$$\pi_{D2} = k_2 \times (\pi_1)^{a2} \times (\pi_2)^{b2} \times (\pi_3)^{c2} \times (\pi_4)^{d2} \times (\pi_5)^{e2}$$

$$(Q) = f2\{(\varphi)(\rho)(T)(S)(mf)\}$$

$$(Q) = 5.27E + 38\{(\varphi)^{0.1698}(\rho)^{-12.5273}(T)^{0.1368}(S)^{0.046}(mf)^{0.5903}\} \qquad (4.18)$$

4.3.6 Model of Dependent Pi Term for π_{D3} (Heat Transfer Coefficient, h)

$$P_3 = W_3 \times X_3$$

−509.553		550	148.0836	1672.324	1909.41	−4028.58	−782.763 K_3
−143.032		148.0836	66.037	451.0388	508.1035	−1087.47	−210.753 a_3
−1549.53	=	1672.324	451.0388	5084.872	5805.529	−12249.4	−2380.06 b_3
−1764.97		1909.41	508.1035	5805.529	6648.779	−13987.7	−2717.48 c_3
3732.924		−4028.58	−1087.47	−12249.4	−13987.7	29513.9	5733.498 d_3
753.4612		−782.763	−210.753	−2380.06	2717.40	5733.498	1164.198 e_3

$$[P3] = [W3] [X3]$$

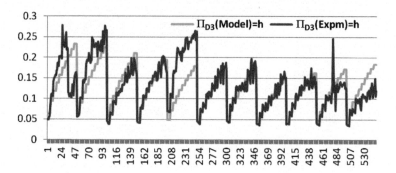

Figure 4.7 Comparison between Model and Experimental Data of Heat Transfer Coefficient, h.

Using MATLAB, $X_3 = W_3/P_3$, after solving the X_3 matrix with K_3 and indices a_3, b_3, c_3, d_3, e_3 are as follows:

K_3	30.7585
a_3	0.1161
b_3	−10.2355
c_3	0.1362
d_3	0.0362
e_3	0.5634

But K_3 is log value. So to convert into normal value taking antilog of Antilog $(30.7585) = 5.73E+30$, Hence the model for dependent term π_{D3} is:

$$\pi_{D3} = K_3 \times (\pi_1)^{a3} \times (\pi_2)^{b3} \times (\pi_3)^{c3} \times (\pi_4)^{d3} \times (\pi_5)^{e3}$$

$$(Q) = f3\{(\varphi)(\rho)(T)(S)(mf)\}$$

$$(Q) = 5.73E + 30\{(\varphi)^{0.1161}(\rho)^{-10.2355}(T)^{0.1362}(S)^{0.0362}(mf)^{0.5634}\} \qquad (4.19)$$

4.4 SAMPLE CALCULATIONS OF PI TERMS

4.4.1 For Two-Wire Method (Table 4.1)

Table 4.1 Sample Readings for Calculations of Multipliers of the L.H.S. and R.H.S. Terms of Equation (4.5) for Formulation of Model for Thermal Conductivity Based on Concentration (K_φ) (π_{D1})

SN	$logZ = Z$	$Log\pi_1 = A$	$Log\pi_2 = B$	$Log\pi_3 = C$	ZA	A	AA	AB	AC
1	−0.25	3.016	2.481	−4	−0.76	3.016	9.097	7.485	−12.1
2	−0.11	3.016	2.481	−0.3	−0.34	3.016	9.097	7.485	−0.91
3	−0.07	3.016	2.481	0	−0.21	3.016	9.097	7.485	0
4	−0.03	3.016	2.481	0.176	−0.1	3.016	9.097	7.485	0.531
5	0.021	3.016	2.481	0.301	0.063	3.016	9.097	7.485	0.908
6	0.064	3.016	2.481	0.398	0.194	3.016	9.097	7.485	1.2

(Continued)

SN	$\log Z = Z$	$\log \pi_1 = A$	$\log \pi_2 = B$	$\log \pi_3 = C$	ZA	A	AA	AB	AC
7	−0.25	3.016	2.489	−4	−0.75	3.016	9.097	7.506	−12.1
8	−0.1	3.016	2.489	−0.3	−0.3	3.016	9.097	7.506	−0.91
9	−0.06	3.016	2.489	0	−0.17	3.016	9.097	7.506	0

ZB	B	AB	BB	BC	ZC	C	AC	BC	CC
−0.62	2.481	7.485	6.158	−9.93	1.004	−4	−12.1	−9.93	16
−0.28	2.481	7.485	6.158	−0.75	0.034	−0.3	−0.91	−0.75	0.091
−0.18	2.481	7.485	6.158	0	0	0	0	0	0
−0.09	2.481	7.485	6.158	0.437	−0.01	0.176	0.531	0.437	0.031
0.052	2.481	7.485	6.158	0.747	0.006	0.301	0.908	0.747	0.091
0.16	2.481	7.485	6.158	0.987	0.026	0.398	1.2	0.987	0.158
−0.62	2.489	7.506	6.193	−9.95	0.996	−4	−12.1	−9.95	16
−0.24	2.489	7.506	6.193	−0.75	0.03	−0.3	−0.91	−0.75	0.091
−0.14	2.489	7.506	6.193	0	0	0	0	0	0

For Conductivity Based on Size (Prob-Sonication Time), K_t (π_{D2})

$\log Z$	$\log \pi_1$	$\log \pi_2$	$\log \pi_3$	ZA	A	AA	AB	AC
0.011	3.016	2.481	1.176	−0.033	3.016	9.097	7.485	3.547
0.024	3.016	2.481	1.477	0.074	3.016	9.097	7.485	4.455
0.055	3.016	2.481	1.653	0.167	3.016	9.097	7.485	4.986
0.087	3.016	2.481	1.778	0.264	3.016	9.097	7.485	5.363
0.006	3.016	2.489	1.176	0.018	3.016	9.097	7.506	3.547
0.043	3.016	2.489	1.477	0.128	3.016	9.097	7.506	4.455
0.076	3.016	2.489	1.653	0.23	3.016	9.097	7.506	4.986
0.107	3.016	2.489	1.778	0.323	3.016	9.097	7.506	5.363
0.017	3.016	2.496	1.176	0.051	3.016	9.097	7.527	3.547
0.055	3.016	2.496	1.477	0.167	3.016	9.097	7.527	4.455

ZB	B	AB	BB	BC	ZC	C	AC	BC	CC
0.027	2.481	7.485	6.158	2.918	−0.013	1.176	3.547	2.918	1.383
0.061	2.481	7.485	6.158	3.665	0.036	1.477	4.455	3.665	2.182
0.137	2.481	7.485	6.158	4.102	0.092	1.653	4.986	4.102	2.733
0.217	2.481	7.485	6.158	4.412	0.155	1.778	5.363	4.412	3.162
0.015	2.489	7.506	6.193	2.927	0.007	1.176	3.547	2.927	1.383
0.106	2.489	7.506	6.193	3.676	0.063	1.477	4.455	3.676	2.182
0.19	2.489	7.506	6.193	4.114	0.126	1.653	4.986	4.114	2.733
0.267	2.489	7.506	6.193	4.425	0.191	1.778	5.363	4.425	3.162
0.043	2.496	7.527	6.228	2.935	0.02	1.176	3.547	2.935	1.383
0.138	2.496	7.527	6.228	3.686	0.082	1.477	4.455	3.686	2.182

(Continued)

Table 4.1 (Continued)

For Thermal Conductivity Based on Shape, K_s (π_{D3})

logZ	Logπ_1	Logπ_2	Logπ_3	ZA	A	AA	AB	AC
0.146	3.016	2.481	1.269	0.441	3.016	9.097	7.485	3.827
0.086	3.024	2.481	1.808	0.261	3.024	9.148	7.505	5.468
−0.004	3.026	2.481	1.153	−0.013	3.026	9.155	7.508	3.489
0.158	3.016	2.489	1.269	0.478	3.016	9.097	7.506	3.827
0.097	3.024	2.489	1.808	0.293	3.024	9.148	7.527	5.468
0.017	3.026	2.489	1.153	0.052	3.026	9.155	7.53	3.489
0.172	3.016	2.496	1.269	0.519	3.016	9.097	7.527	3.827
0.114	3.024	2.496	1.808	0.345	3.024	9.148	7.548	5.468
0.033	3.026	2.496	1.153	0.101	3.026	9.155	7.551	3.489
0.185	3.016	2.502	1.269	0.559	3.016	9.097	7.548	3.827

ZB	B	AB	BB	BC	ZC	C	AC	BC	CC
0.363	2.481	7.485	6.158	3.148	0.185	1.269	3.827	3.148	1.61
0.214	2.481	7.505	6.158	4.487	0.156	1.808	5.468	4.487	3.269
−0.011	2.481	7.508	6.158	2.862	−0.005	1.153	3.489	2.862	1.33
0.394	2.489	7.506	6.193	3.158	0.201	1.269	3.827	3.158	1.61
0.241	2.489	7.527	6.193	4.499	0.175	1.808	5.468	4.499	3.269
0.042	2.489	7.53	6.193	2.87	0.02	1.153	3.489	2.87	1.33
0.429	2.496	7.527	6.228	3.166	0.218	1.269	3.827	3.166	1.61
0.284	2.496	7.548	6.228	4.512	0.206	1.808	5.468	4.512	3.269
0.083	2.496	7.551	6.228	2.878	0.039	1.153	3.489	2.878	1.33
0.464	2.502	7.548	6.262	3.175	0.235	1.269	3.827	3.175	1.61

4.4.2 For Experimental Model (Radiator as a Heat Exchanger [Tables 4.2 and 4.3])

Table 4.2 Sample Readings for Calculations of Multipliers of the R.H.S. Terms of Equation (4.15) for Formulation of Model (for Temperature Difference, Heat Flow and Heat Transfer Coefficient)

logZ	Logπ_1	Logπ_2	Logπ_3	Logπ_4	Logπ_5
Z	A	B	C	D	E
1.0969	−0.301	3.0257	3.5563	−7.285	−2.079
1.0934	−0.301	3.0257	3.5563	−7.285	−2.079
1.0212	−0.301	3.0257	3.5563	−7.285	−2.079
1.0607	−0.301	3.0257	3.5563	−7.285	−2.079
1.0294	−0.301	3.0257	3.5563	−7.285	−2.079
1.1038	−0.301	3.0257	3.5563	−7.285	−1.778
1.1303	−0.301	3.0257	3.5563	−7.285	−1.778
1.1173	−0.301	3.0257	3.5563	−7.285	−1.778
1.0755	−0.301	3.0257	3.5563	−7.285	−1.778
1.0253	−0.301	3.0257	3.5563	−7.285	−1.778

(Continued)

A	AA	AB	AC	AD	AE
-0.301	0.0906	-0.911	-1.071	2.193	0.6259
-0.301	0.0906	-0.911	-1.071	2.193	0.6259
-0.301	0.0906	-0.911	-1.071	2.193	0.6259
-0.301	0.0906	-0.911	-1.071	2.193	0.6259
-0.301	0.0906	-0.911	-1.071	2.193	0.6259
-0.301	0.0906	-0.911	-1.071	2.193	0.5353
-0.301	0.0906	-0.911	-1.071	2.193	0.5353
-0.301	0.0906	-0.911	-1.071	2.193	0.5353
-0.301	0.0906	-0.911	-1.071	2.193	0.5353
-0.301	0.0906	-0.911	-1.071	2.193	0.5353

B	AB	BB	BC	BD	BE
3.0257	-0.911	9.155	10.76	-22.04	-6.291
3.0257	-0.911	9.155	10.76	-22.04	-6.291
3.0257	-0.911	9.155	10.76	-22.04	-6.291
3.0257	-0.911	9.155	10.76	-22.04	-6.291
3.0257	-0.911	9.155	10.76	-22.04	-6.291
3.0257	-0.911	9.155	10.76	-22.04	-5.38
3.0257	-0.911	9.155	10.76	-22.04	-5.38
3.0257	-0.911	9.155	10.76	-22.04	-5.38
3.0257	-0.911	9.155	10.76	-22.04	-5.38
3.0257	-0.911	9.155	10.76	-22.04	-5.38

C	AC	BC	CC	CD	CE
3.5563	-1.071	10.76	12.647	-25.91	-7.394
3.5563	-1.071	10.76	12.647	-25.91	-7.394
3.5563	-1.071	10.76	12.647	-25.91	-7.394
3.5563	-1.071	10.76	12.647	-25.91	-7.394
3.5563	-1.071	10.76	12.647	-25.91	-7.394
3.5563	-1.071	10.76	12.647	-25.91	-6.324
3.5563	-1.071	10.76	12.647	-25.91	-6.324
3.5563	-1.071	10.76	12.647	-25.91	-6.324
3.5563	-1.071	10.76	12.647	-25.91	-6.324
3.5563	-1.071	10.76	12.647	-25.91	-6.324

D	AD	BD	CD	DD	DE
-7.285	2.193	-22.04	-25.91	53.072	15.147
-7.285	2.193	-22.04	-25.91	53.072	15.147
-7.285	2.193	-22.04	-25.91	53.072	15.147
-7.285	2.193	-22.04	-25.91	53.072	15.147
-7.285	2.193	-22.04	-25.91	53.072	15.147

(Continued)

Table 4.2 (Continued)

D	AD	BD	CD	DD	DE
−7.285	2.193	−22.04	−25.91	53.072	12.954
−7.285	2.193	−22.04	−25.91	53.072	12.954
−7.285	2.193	−22.04	−25.91	53.072	12.954
−7.285	2.193	−22.04	−25.91	53.072	12.954
−7.285	2.193	−22.04	−25.91	53.072	12.954

ZE	E	AE	BE	CE	DE	EE
−2.281	−2.079	0.6259	−6.291	−7.394	15.147	4.323
−2.273	−2.079	0.6259	−6.291	−7.394	15.147	4.323
−2.123	−2.079	0.6259	−6.291	−7.394	15.147	4.323
−2.205	−2.079	0.6259	−6.291	−7.394	15.147	4.323
−2.14	−2.079	0.6259	−6.291	−7.394	15.147	4.323
−1.963	−1.778	0.5353	−5.38	−6.324	12.954	3.1618
−2.01	−1.778	0.5353	−5.38	−6.324	12.954	3.1618
−1.987	−1.778	0.5353	−5.38	−6.324	12.954	3.1618
−1.912	−1.778	0.5353	−5.38	−6.324	12.954	3.1618
−1.823	−1.778	0.5353	−5.38	−6.324	12.954	3.1618

Table 4.3 Sample Readings for Calculations of Multipliers of the L.H.S. Terms of Equation 15 for Formulation of Model for Temperature Difference Model, ΔT (π_{D1})

logZ	ZA	ZB	ZC	ZD	ZE
1.0969	−0.33	3.3189	3.9009	−7.991	−2.281
1.0934	−0.329	3.3084	3.8885	−7.966	−2.273
1.0212	−0.307	3.0898	3.6317	−7.439	−2.123
1.0607	−0.319	3.2094	3.7722	−7.727	−2.205
1.0294	−0.31	3.1146	3.6608	−7.499	−2.14
1.1038	−0.332	3.3398	3.9255	−8.041	−1.963
1.1303	−0.34	3.4201	4.0198	−8.235	−2.01
1.1173	−0.336	3.3805	3.9734	−8.139	−1.987
1.0755	−0.324	3.2543	3.825	−7.835	−1.912
1.0253	−0.309	3.1023	3.6463	−7.469	−1.823

For Heat Flow, Q (π_{D2})

logZ	ZA	ZB	ZC	ZD	ZE
−0.36949	0.111227	−1.11796	−1.31401	2.691746	0.768231
−0.37298	0.112277	−1.12852	−1.32641	2.717159	0.775484
−0.44521	0.134021	−1.34707	−1.58329	3.243378	0.925668
−0.4057	0.122128	−1.22753	−1.44279	2.955555	0.843523
−0.43701	0.131554	−1.32228	−1.55415	3.183681	0.908631
−0.06156	0.018533	−0.18627	−0.21894	0.448497	0.10947
−0.03503	0.010546	−0.106	−0.12459	0.255223	0.062295

(Continued)

logZ	ZA	ZB	ZC	ZD	ZE
−0.0481	0.014478	−0.14553	−0.17104	0.350384	0.085522
−0.08982	0.027039	−0.27177	−0.31943	0.654349	0.159714
−0.14006	0.042163	−0.42379	−0.4981	1.02036	0.249051

For Heat Transfer Coefficient, h (π_{D3})

logZ	ZA	ZB	ZC	ZD	ZE
−1.30644	0.393276	−3.9529	−4.64608	9.51749	2.716316
−1.29458	0.389706	−3.91702	−4.6039	9.431093	2.691658
−1.31015	0.394395	−3.96415	−4.6593	9.544576	2.724046
−1.258	0.378697	−3.80636	−4.47384	9.164659	2.615617
−1.2409	0.37355	−3.75462	−4.41303	9.040095	2.580066
−1.01041	0.304164	−3.05722	−3.59333	7.360927	1.796663
−0.95663	0.287975	−2.8945	−3.40208	6.969157	1.70104
−0.95053	0.286139	−2.87604	−3.38038	6.924713	1.690192
−0.96183	0.289539	−2.91022	−3.42055	7.006993	1.710275
−0.97364	0.293096	−2.94597	−3.46257	7.093069	1.731284

Chapter 5

Analysis using SPSS Statistical Packages Software

5.1 INTRODUCTION

One of the most popular statistical packages, SPSS can perform highly complex data analysis with simple instructions. SPSS is capable of handling large amount of data. In this study, descriptive statistics (arithmetic mean, standard deviation, maximum and minimum value of variables, etc.), data testing (normality test, data adequacy, reliability and validity) and final analysis (internal consistency, factor analysis, analysis of variance, multiple regression analysis and hypothesis testing) are carried out through SPSS software version 20.0.

1. *Linear regression*: Linear regression is used to specify the nature of the relation between two variables. Another way of looking at it is as follows: given the value of one variable (called the independent variable in SPSS), how can you predict the value of some other variable (called the dependent variable in SPSS)?
2. *Descriptive statistics*: The Descriptive Statistics part of the output gives the mean, standard deviation and observation count (N) for each of the dependent and independent variables.
3. *Correlations*: The Correlations part of the output shows the correlation coefficients. This output is organized differently from the output from the correlation procedure. The first row gives the correlations between the independent and dependent variables.
4. *Variables Entered/Removed*: The Variables Entered/Removed part of the output simply states which independent variables are part of the equation (extravert in this example) and what is the dependent variable.
5. *Model Summary*: The Model Summary part of the output is most useful when performing multiple regression (which is not done here). Capital R is the multiple correlation coefficient that tells us how strongly the multiple independent variables are related to the dependent variable. In the simple bivariate case (what we are doing) R = | r | (multiple correlation equals the absolute value of the bivariate correlation.) R^2 is useful as it gives us the coefficient of determinant.

DOI: 10.1201/9781003432111-5

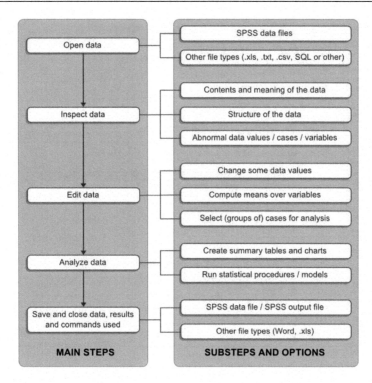

Figure 5.1 SPSS Project Workflow.

6. *ANOVA*: The ANOVA part of the output is not very useful for our purposes. It basically tells us whether the regression equation is explaining a statistically significant portion of the variability in the dependent variable from variability in the independent variables.

7. *Coefficients part*: The Coefficients part of the output gives us the values that we need in order to write the regression equation. The regression equation will take the form: *Predicted variable (dependent variable) = slope * independent variable + intercept.* A slope of 0 is a horizontal line, a slope of 1 is a diagonal line from the lower left to the upper right, and a vertical line has an infinite slope. The intercept is where the regression line strikes the Y axis when the independent variable has a value of 0.

5.2 DEVELOPING THE SPSS MODEL FOR INDIVIDUAL PI TERMS

Here five independent pi terms (i.e., π_1, π_2, π_3, π_4, π_5) and three dependent pi terms (π_{D1}, π_{D2}, π_{D3}) have been identified in the design of experimentation and are available for the model formulation in radiator experimental setup.

Independent π terms = (π_1, π_2, π_3, π_4, π_5) and dependent π terms = (π_{D1}, π_{D2}, π_{D3})

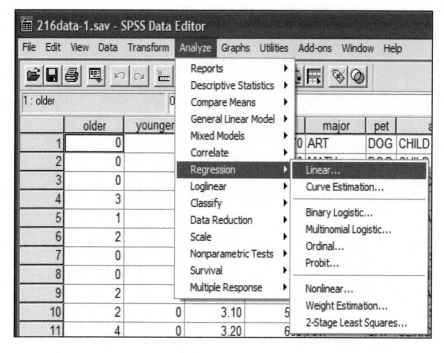

Figure 5.2 Linear Regression Command at Analyze | Regression | Linear.

Each dependent π term is assumed to be a function of the available independent π terms. By using SPSS software version 20.0, linear regression is carried out. Linear regression is used to specify the nature of the relation between two variables. Another way of looking at it is, given the value of one variable (called the independent variable in SPSS), how can you predict the value of some other variable (called the dependent variable in SPSS)?

5.3 SPSS OUTPUT FOR THERMAL CONDUCTIVITY K_ϕ (CONCENTRATION)

Dependent Variable: π_{D1}

Descriptive Statistics

	Mean	Std. Deviation	N
π_{D1}	.8414	.21574	330
π_1	4832.0000	1162.70056	330
π_2	328.0000	15.83540	330
π_3	1.2500	.85518	330

(Continued)

Correlations

		π_{DI}	π_1	π_2	π_3
Pearson	π_{DI}	1.000	−.154	.316	.747
Correlation	π_1	−.154	1.000	.000	.000
	π_2	.316	.000	1.000	.000
	π_3	.747	.000	.000	1.000
Sig. (1-tailed)	π_{DI}	.	.003	.000	.000
	π_1	.003	.	.500	.500
	π_2	.000	.500	.	.500
	π_3	.000	.500	.500	.
N	π_{DI}	330	330	330	330
	π_1	330	330	330	330
	π_2	330	330	330	330
	π_3	330	330	330	330

Variables Entered/Removed[a]

Model	Variables Entered	Variables Removed	Method
1	π_3, π_2, π_1[b]	.	Enter

a. Dependent Variable: π_{D1} b. All requested variables entered

Model Summary[b]

Model	R	R^2	Adjusted R^2	Std. Error of the Estimate
1	.826[a]	.682	.679	.12219

a. Predictors: (Constant), π_3, π_2, π_1 b. Dependent Variable: π_{D1}

ANOVA[a]

Model		Sum of Squares	df	Mean Square	F	Sig.
1	Regression	10.446	3	3.482	233.234	.000[b]
	Residual	4.867	326	.015		
	Total	15.313	329			

(Continued)

a. Predictors: (Constant), π_3, π_2, π_1 b. Dependent Variable: π_{D1}

Coefficients^a

Model	Unstandardized Coefficients		Standardized Coefficients			95.0% Confidence Interval for B	
	B	Std. Error	Beta	t	Sig.	Lower Bound	Upper Bound
I (Constant)	−.669	.143		−4.683	.000	−.950	−.388
Pi1	−2 854E-005	.000	−.154	−4.926	.000	.000	.000
Pi2	.004	.000	.316	10.123	.000	.003	.005
Pi3	.139	.003	.747	23.937	.000	.173	.204

^{a} Dependent Variable: Pi01

Model Equation:

$$K\varphi = -0.669 - 2.854 \times 10^{-5}\ (\pi_1) + 0.004\ (\pi_2) + 0.189\ (\pi_3)$$

$$K\varphi = -0.669 - 2.854 \times 10^{-5}\ (\rho) + 0.004\ (T) + 0.189\ (\varphi)$$

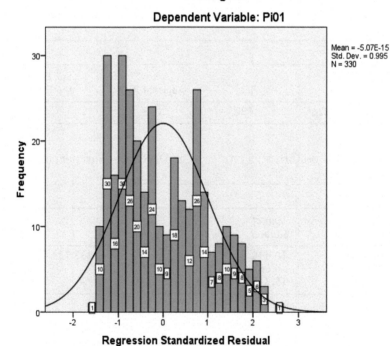

Figure 5.3 Graph for SPSS Model for $\pi_{D1} = K_\Phi$ (concentration) for Histogram and Normal P-P Plot of Regression Standardized Residual.

Normal P-P Plot of Regression Standardized Residual

Dependent Variable: Pi01

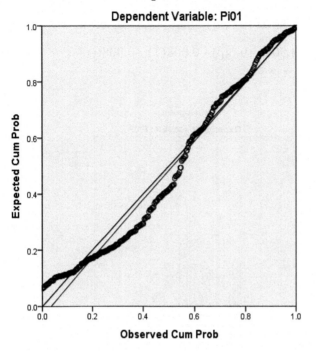

R^2 Linear = 0.976

Figure 5.3 (Continued)

5.4 SPSS OUTPUT FOR THERMAL CONDUCTIVITY K_T (SIZE)

Dependent Variable: π_{D2}

Coefficients[a]

Model	Unstandardized Coefficients E	Std. Error	Standardized Coefficients Beta	t	Sig.	95.0% Confidence Interval for B Lower Bound	Upper Bound
1 (Constant)	−1.716	.147		−11.662	.000	−2.005	−1.426
Pi1	−4.807E-005	.000	−.255	−8.080	.000	.000	.000
Pi2	.008	.000	.615	19.445	.000	.008	.009
Pi3	.008	.000	.584	18.483	.000	.007	.008

[a] Dependent Variable: Pi02

Model Equation:

$$K_t = -1.716 - 4.807 \times 10^{-5}(\pi_1) + 0.008(\pi_2) + 0.008(\pi_3)$$

$$K_t = -1.716 - 4.807 \times 10^{-5}(\rho) + 0.008(T) + 0.008(t)$$

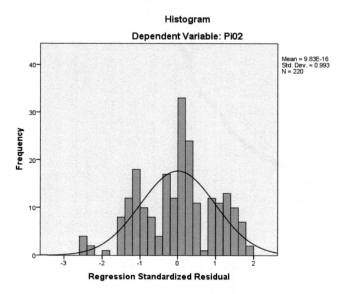

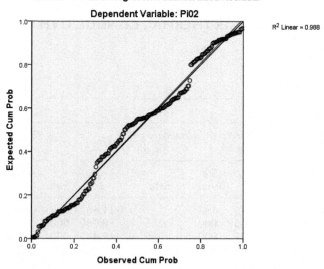

Figure 5.4 Graph for SPSS Model for $\pi_{D2} = K_t$ (Size) for Histogram and Normal P-P Plot of Regression Standardized Residual+.

5.5 SPSS OUTPUT FOR THERMAL CONDUCTIVITY BASED ON SHAPE

Dependent Variable: π_{D3}

Coefficients^a

Model	Unstandardized Coefficients		Standardized Coefficients		
	E	Std. Error	Beta	t	Sig.
I (Constant)	1.180	.570		2.072	.040
Pi I	-.002	.000	-.349	-5.177	.000
Pi2	.006	.001	.359	5.327	.000
Pi3	.002	.001	.132	1.958	.052

^{a} Dependent Variable: Pi03 K_s

Model Equation:

$$K_s = 1.180 - 0.002\,(\pi_1) + 0.006\,(\pi_2) + 0.002\,(\pi_3)$$

$$K_s = 1.180 - 0.002\,(\rho) + 0.006\,(T) + 0.002\,(s)$$

5.6 SPSS OUTPUT FOR π_{D1} (TEMPERATURE DIFFERENCE, ΔT)

Coefficients^a

Model	Unstandardized Coefficients		Standardized Coefficients		
	B	Std. Error	Beta	t	Sig.
I (Constant)	53.165	6.465		8.224	.000
Pi2	-.041	.006	-.228	-7.253	.000
Pi3	.000	.000	.140	4.538	.000
Pi4	19262112.43	10695390.73	.055	1.801	.072
Pi5	-83.773	3.510	-.680	-23.864	.000

^{a} Dependent Variable: π_{D1} (ΔT)

Model Equation:

$$(\Delta T) = 1.57E-08\left\{(\varphi)^{0.2028}\,(\rho)^{-12.5534}\,(T)^{0.1366}\,(S)^{0.0457}\,(m_f)^{-0.4097}\right\}$$

Histogram
Dependent Variable: π_{D1} ΔT

Normal P-P Plot
Dependent Variable: π_{D1} ΔT

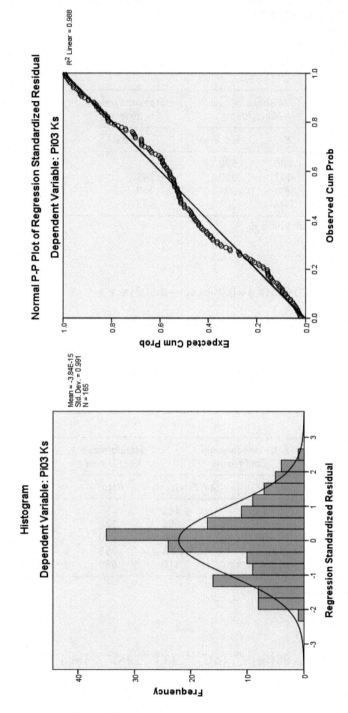

Figure 5.5 Graph for SPSS Model for $\pi_{D3} = K_s$ (Shape) for Histogram and Normal P-P Plot of Regression Standardized Residual.

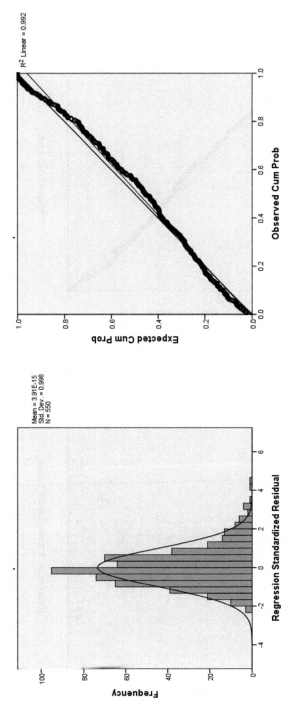

Figure 5.5 Graph for SPSS Model for πD I (Temperature Difference, ΔT) for Histogram and Normal P-P Plot of Regression Standardized Residual.

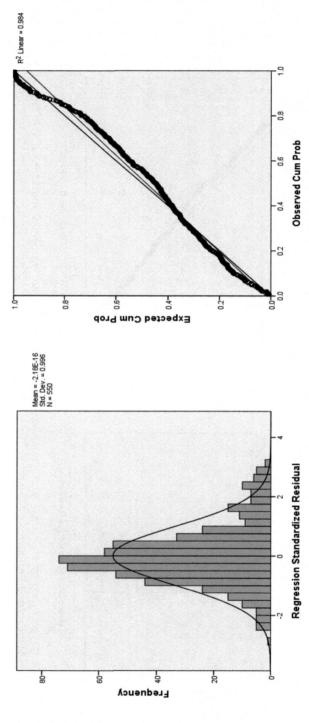

Figure 5.7 Graph for SPSS Model for π_{D2} (Heat Flow, Q) for Histogram and Normal P-P Plot of Regression Standardized Residual.

5.7 SPSS OUTPUT FOR Π_{D2} (HEAT FLOW, Q)

Coefficients[a]

Model	Unstandardized Coefficients		Standardized Coefficients		
	B	Std. Error	Beta	t	Sig.
I (Constant)	7.865	1.125		6.993	.000
Pi2	−.007	.001	−.238	−7.214	.000
Pi3	6.275E-005	.000	.124	3.848	.000
Pi4	2435569.848	1860641.145	.042	1.309	.191
Pi5	13.309	.611	.650	21.794	.000

[a] Dependent Variable: π_{D2} (Q)

Model Equation:

$$(Q) = 5.27E + 38\left\{(\varphi)^{0.1698}(\rho)^{-12.5273}(T)^{0.1368}(S)^{0.046}(m_f)^{0.5903}\right\}$$

Histogram
Dependent Variable: π_{D2} Q

Normal P-P Plot
Dependent Variable: π_{D2} Q

5.8 SPSS OUTPUT FOR Π_{D3} (HEAT TRANSFER COEFFICIENT, H)

Coefficients[a]

Model	Unstandardized Coefficients		Standardized Coefficients		
	B	Std. Error	Beta	t	Sig.
I (Constant)	.934	.126		7.421	.000
Pi2	−.001	.000	−.242	−7.587	.000
Pi3	7.263E-006	.000	.124	3.980	.000
Pi4	237529.168	208185.306	.036	1.141	.254
Pi5	1.581	.068	.670	23.141	.000

[a] Dependent Variable: π_{D3} (h)

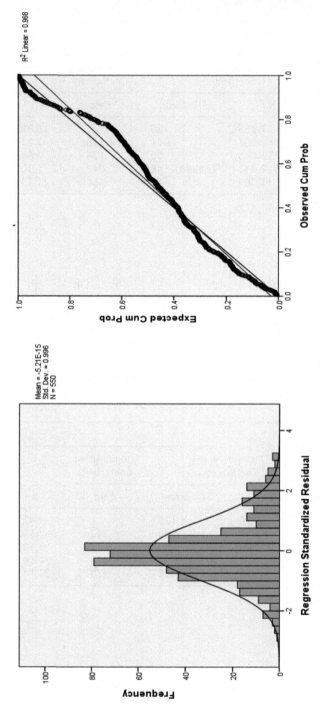

Figure 5.8 Graph for SPSS Model for π_{D3} (Heat Transfer Coefficient, h) for Histogram and Normal P-P Plot of Regression Standardized Residual.

Model Equation:

$$(h) = 5.73E + 30\left\{(\varphi)^{0.1161}(\rho)^{-10.2355}(T)^{0.1362}(S)^{0.0362}(m_f)^{0.5634}\right\}$$

Histogram   Normal P-P Plot
Dependent Variable: π_{D3} h   Dependent Variable: π_{D3} h

The model summary shows the values of R, R^2, adjusted R^2 and standard error of linear regression by using SPSS software are much better.

Chapter 6

Analysis of Model using Artificial Neural Network Programming

6.1 INTRODUCTION

An artificial neural network (ANN) consists of three layers – the input layer, the hidden layer and the output layer. Its nodes represent neurons of the brain. The three neurons are interconnected with nodes like that of neurons in the brain. The specific mapping performed depends upon the architecture and synaptic weight values between the neurons of an ANN. An artificial neural network consists of highly distributed representation and transformation that work in parallel. The control is distributed by highly interconnected numerous nodes. ANN is developed on a concept similar to a black box. ANN is trained itself with the help of input and output data, like a human brain learning with reception of similar stimuli. As ANN trains itself within input and output data, it usually operates without a prior theory that guides or restricts a relation between output and input. Ultimately the accuracy of predicted output rather than a specific description of paths and the relation between input and output is the eventual goal of ANN model. The input data are preprocessed and passed through the hidden layer nodes.

A transfer function which is nonlinear in nature assigns weights to the information as it passes through hidden layer nodes, mimicking the transfer of information as it passes through the synapses of a brain. The role of modelling using ANN is to develop responses by assigning the weights in such a way that it represents the true relationship that exists between output and input. Interpolation of function takes place between input and output neuron layers during training of the network. ANN does not give an explicit description of function but gives the pattern recognition.

6.2 PROCEDURE FOR ARTIFICIAL NEURAL NETWORK PHENOMENON

Different software/tools have been developed to construct ANN. MATLAB®, an internationally accepted tool, has been selected for developing ANN for

DOI: 10.1201/9781003432111-6

the complex phenomenon. The various steps followed in developing the algorithm from ANN are as follows:

1. The observed data from the experimentation is separated into two parts – input data or the data of independent pi terms, and output data or the data of dependent pi terms. The input data and output data are imported to the program respectively.
2. The input and output data are read by function and appropriately sized.
3. In the preprocessing step, the input and output data are normalized using mean and standard deviation.
4. Through principal component analysis, the normalized data are uncorrelated. This is achieved by using the "prestd" function. The input and output data are then categorized in three categories – testing, validation and training. The common practice is to select the initial 25% training, the last 25% of data for validation and the middle overlapping 50% of data for testing.
5. The data are then stored in structures for training, testing and validation.
6. Looking at the pattern of the data feed, a forward back propagation type neural network is chosen.
7. This network is then trained using the training data. The computational errors in the actual and target data are computed, and then the network is simulated.
8. The uncorrelated output data are again transformed into the original form by using the "poststd" function.
9. After simulating the ANN, it is found that experimentally observed values are very close and in good agreement with the ANN predicted values.

Figure 6.1 shows a simple multilayer feed forward network for ANN, and Figure 6.2 shows the flow diagram of ANN simulation.

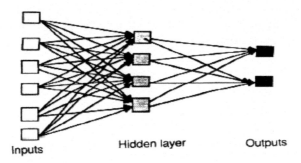

Figure 6.1 Simple multilayer feed forward network (ANN).

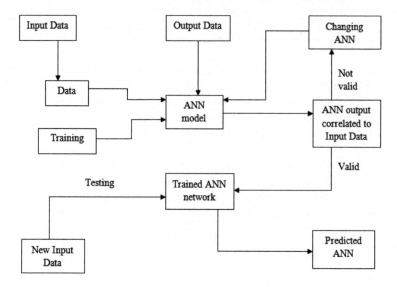

Figure 6.2 ANN simulation flow diagram.

6.3 PERFORMANCE OF MODELS BY ANN

The models have been formulated mathematically as well as by using ANN. Now we have three sets of values of three dependent pi terms of the two-wire method for thermal conductivity calculation – values computed by experimental observations, values computed by mathematical model and values obtained by ANN. All these values match very well with each other. This is justified by calculating their respective mean values and standard errors of estimation. The comparison of values of dependent pi terms obtained by experimentation, mathematical models are calculated. From the values of "standard error estimation" and "percent error" is calculated, it seems that the mathematical models and the ANN developed using MATLAB can be successfully used for computation of dependent terms for a given set of the independent pi terms. The values of R^2 error in ANN, number of iterations, values of the regression coefficients for dependent pi terms are calculated. The values are fairly good.

6.3.1 ANN using SPSS o/p for Thermal Conductivity Kϕ

Case Processing Summary

		N	Percent
	Training	227	68.8%
Sample	Testing	65	19.7%
	Holdout	38	11.5%
Valid		330	100.0%
Excluded		0	
Total		330	

Network Information

Input Layer	Covariates	1	π_1
		2	π_2
		3	π_3
	Number of Units[a]		3
	Rescaling Method for Covariates		Standardized
Hidden Layer(s)	Number of Hidden Layers		1
	Number of Units in Hidden Layer 1[a]		2
	Activation Function		Hyperbolic tangent
Output Layer	Dependent Variables	1	π_{DI}
	Number of Units		1
	Rescaling Method for Scale Dependents		Standardized
	Activation Function		Identity
	Error Function		Sum of Squares

a. Excluding the bias unit

Model Summary

Training	Sum of Squares Error	38.121
	Relative Error	.337
	Stopping Rule Used	One consecutive step(s) with no decrease in error[a]
	Training Time	0:00:00.08
Testing	Sum of Squares Error	11.053
	Relative Error	.245
Holdout	Relative Error	.219

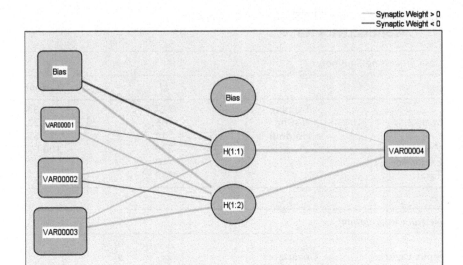

Hidden layer activation function: Hyperbolic tangent

Output layer activation function: Identity

Figure 6.3 ANN topology (network diagram) for concentration K_Φ (π_{D1}).

Dependent Variable: π_{D1} (Kφ)

a. Error computations are based on the testing sample.

Parameter Estimates

| | | Predicted | | |
| | | Hidden Layer I | | Output Layer |
Predictor		H(1:1)	H(1:2)	VAR00004
	(Bias)	−.317	.836	
Input Layer	VAR00001	−.173	.307	
	VAR00002	.268	−.254	
	VAR00003	.306	.447	
	(Bias)			.065
Hidden Layer I	H(1:1)			2.104
	H(1:2)			.834

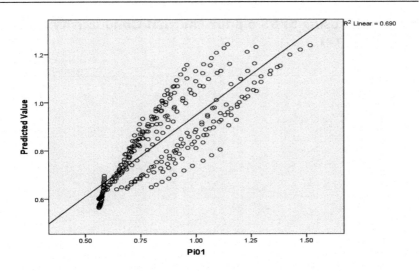

Predicted by Observed Chart

Independent Variable Importance		
	Importance	Normalized Importance
π_1	.106	17.7%
π_2	.295	49.2%
π_3	.599	100.0%

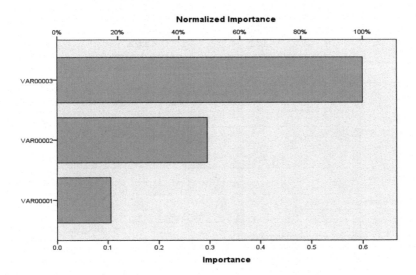

Figure 6.4 Chart of independent variable importance for concentration.

6.3.2 ANN using SPSS o/p for Thermal Conductivity K_t (Size)

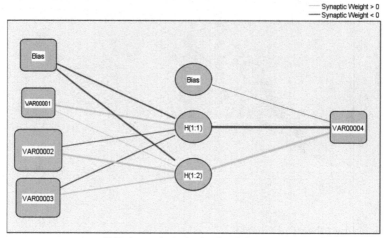

Hidden layer activation function: Hyperbolic tangent

Output layer activation function: Identity

Figure 6.5 ANN topology (network diagram) for size K_t (π_{D2}).

Independent Variable Importance		
	Importance	Normalized Importance
π_1	.168	37.3%
π_2	.450	100.0%
π_3	.382	85.0%

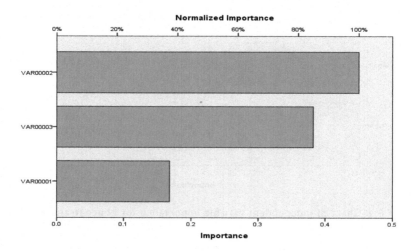

Figure 6.6 Chart of independent variable importance for size K_t (π_{D2}).

6.3.3 ANN using SPSS o/p for Thermal Conductivity K$_s$ (Shape)

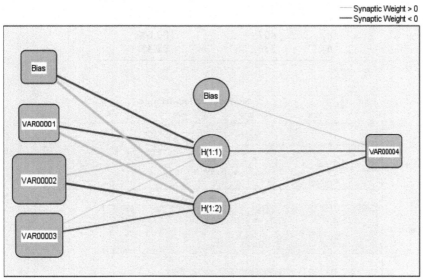

Hidden layer activation function: Hyperbolic tangent

Output layer activation function: Identity

Figure 6.7 ANN topology (network diagram) for shape K$_s$ (πD3).

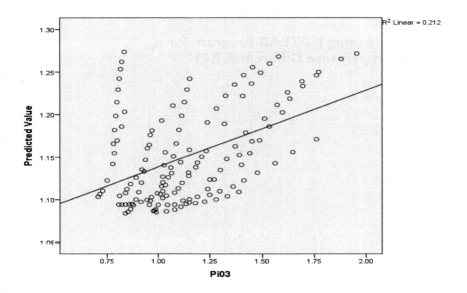

Independent Variable Importance		
	Importance	Normalized Importance
π_1	.257	63.2%
π_2	.407	100.0%
π_3	.336	82.4%

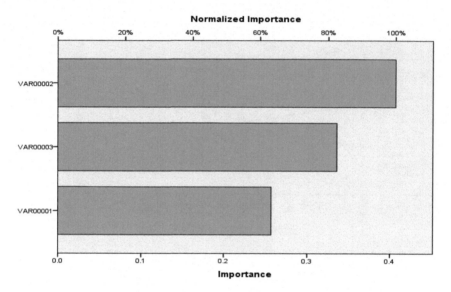

Figure 6.8 Chart of independent variable importance for shape K_s (π_{D3}).

6.3.4 ANN using MATLAB Program for π_{D1} (Temperature Difference, ΔT)

```
clear all;
close all;
inputs3=[]
a1=inputs3
a2=a1
input_data=a2;
output3=[]
y1=output3
y2=y1
size(a2);
size(y2);
p=a2';
sizep=size(p);
```

```
t=y2';
sizet=size(t);
[S Q]=size(t)
[pn,meanp,stdp,tn,meant,stdt] = prestd(p,t);
net = newff(minmax(pn),[550 1],{'logsig'
  'purelin'},'trainlm');
net.performFcn='mse';
net.trainParam.goal=.99; net.trainParam.show=50;
  net.trainParam.epochs=50; net.trainParam.mc=0.05;
net = train(net,pn,tn);
an = sim(net,pn);
[a] = poststd(an,meant,stdt);
error=t-a;
x1=1:550;
plot(x1,t,'rs-',x1,a,'b-')
legend('Experimental','Neural');
title('Output (Red) and Neural Network Prediction
  (Blue) Plot'); xlabel('Experiment No.');
ylabel('Output');
grid on;
figure
error_percentage=100*error./t
plot(x1,error_percentage)
legend('percentage error');
axis([0 550-100 100]);
title('Percentage Error Plot in Neural Network
  Prediction');
xlabel('Experiment No.');
ylabel('Error in %');
grid on;
for ii=1:550
xx1=input_data(ii,1);
yy2=input_data(ii,2);
zz3=input_data(ii,3); xx4=input_data(ii,4);
yy5=input_data(ii,5);
pause
yyy(1,ii)= 1.57E+38*power(xx1,
  0.2028)*power(yy2, -12.5534)*power(zz3,
  0.1366)*power(xx4, 0.0457)*power(yy5, -0.4097);
  yy_practical(ii)=(y2(ii,1));
yy_eqn(ii)=(yyy(1,ii))
yy_neur(ii)=(a(1,ii))
yy_practical_abs(ii)=(y2(ii,1));
yy_eqn_abs(ii)=(yyy(1,ii));
yy_neur_abs(ii)=(a(1,ii));
```

```
pause
end figure;
plot(x1,yy_practical_abs,'r-',x1,yy_
  eqn_abs,'b-',x1,yy_neur_abs,'k-');
  legend('Experimental','Math.
  Model','Neural(ANN)');
title('Comparison between practical data, equation
  based data and neural based data');
xlabel('Experimental');
grid on;
figure;
plot(x1,yy_practical_abs,'r-',x1,yy_eqn_abs,'b-');
  legend('Practical',' Equation');
title('Comparison between practical data, equa-
  tion based data and neural based data');
  xlabel('Experimental');
grid on; figure;
plot(x1,yy_practical_abs,'r-',x1,yy_neur_abs,'k-');
  legend('Practical','Neural');
title('Comparison between practical data, equation
  based data and neural based data'); xlabel
  ('Experimental');
grid on;
error1=yy_practical_abs-yy_eqn_abs
figure
error_percentage1=100*error1./yy_practical_abs;
plot(x1,error_percentage,'k-',x1,error_percentage1,
  'b-');
legend('Neural','Equation');
axis([0 550-100 100]);
title('Percentage Error Plot in Equation (blue),
  Neural Network (black) Prediction');
xlabel('Experiment No.');
ylabel('Error in %');
grid on;
meanexp=mean(output3)
meanann=mean(a)
meanmath=mean(yy_eqn_abs)
mean_absolute_error_performance_func-
  tion = mae(error) mean_squared_error_performance_
  function = mse(error)
net = newff(minmax(pn),[550 1],{'logsig' 'purelin'},'
  trainlm','learngdm','msereg');
```

```
an = sim(net,pn);
[a] = poststd(an,meant,stdt);
error=t(1,[1:550])-a(1,[1:550]);
net.performParam.ratio = 20/(20+1);
perf = msereg(error,net) rand('seed',1.818490882E9)
[ps] = minmax(p);
[ts] = minmax(t);
numInputs = size(p,1);
numHiddenNeurons = 550;
numOutputs = size(t,1);
net = newff(minmax(p), [numHiddenNeurons,numOutp
   uts]);
[pn,meanp,stdp,tn,meant,stdt] = prestd(p,t);
[ptrans,transmit]=prepca(pn,0.001);
[R Q]=size(ptrans);
testSamples= 200:1:Q;
validateSamples=200:1:Q;
trainSamples= 150:1:Q;
validation.P=ptrans(:,validateSamples);
validation.T=tn(:,validateSamples);
testing.P= ptrans(:,testSamples);
testing.T= tn(:,testSamples)
ptr= ptrans(:,trainSamples);
ttr= tn(:,trainSamples);
net = newff(minmax(ptr),[550 1],{'logsig'
   'purelin'},'trainlm');
[net,tr] = train(net,ptr,ttr,[],[],validation,test
   ing);
plot(tr.epoch,tr.perf, 'r',tr.epoch,tr.vperf,
   'g',tr.epoch,tr.tperf, 'h');
legend('Training', 'validation', 'Testing',-1);
ylabel('Error');
an=sim(net,ptrans); a=poststd(an,meant,stdt);
   pause;
figure
[m,b,r] = postreg(a,t);
```

6.3.5 Comparison of Various Model Values

The estimated values of all forms of the model and its comparison are calculated. The results shown in these tables come from using the two-wire method

Table 6.1 Comparison of Various Model Values for Thermal Conductivity Based on Concentration **(K$_\Phi$) (π_{DI})**

S.N.	K (Expem)	K (Math. Model)	K (SPSS_ ANN)	Z = K (Linear Model)	%Error betn Expm & Math	%Error betn Expm & ANN	%Error betn Expm & SPSS
1	0.5612	0.50695	0.57	0.52494	9.6667	1.568068	6.461155
2	0.772	0.764937	0.64	0.61919	0.914869	17.09845	19.79404
3	0.85	0.79098	0.71	0.71347	6.943522	16.47059	16.06235
4	0.924	0.806623	0.79	0.80774	12.70311	14.50216	12.58225
5	1.049	0.81791	0.87	0.90202	22.02959	17.06387	14.01144
6	1.16	0.826773	0.95	0.99629	28.72651	18.10345	14.11293
7	0.5637	0.520097	0.57	0.54647	7.735215	1.117616	3.05659
8	0.798	0.784773	0.65	0.64072	1.657472	18.54637	19.70927
9	0.876	0.811492	0.73	0.735	7.363977	16.66667	16.09589
10	0.963	0.82754	0.81	0.82927	14.06642	15.88785	13.88681
:	:	:	:	:	:	:	:
:	:	:	:	:	:	:	:
321	0.933	0.928402	0.85	0.84274	0.492807	8.896034	9.674169
322	1.053	0.946763	0.95	0.93702	10.08897	9.781576	11.01425
323	1.129	0.96001	1.04	1.03129	14.96809	7.883082	8.654562
324	1.236	0.970413	1.13	1.12557	21.48762	8.576052	8.934466
325	0.5862	0.608453	0.66	0.67574	3.79615	12.58956	15.27465
326	0.898	0.918094	0.76	0.77	2.237678	15.36748	14.2539
327	0.96	0.949352	0.87	0.86427	1.109214	9.375	9.971875
328	1.084	0.968127	0.98	0.95855	10.6894	9.594096	11.57288
329	1.16	0.981673	1.08	1.05283	15.37302	6.896552	9.238793
330	1.267	0.99231	1.16	1.1471	21.68031	8.445146	9.463299
Avg	**0.841378**	**0.82899**	**0.837727**	**0.841378**	**1.472324**	**0.433866**	**3.6E-06**

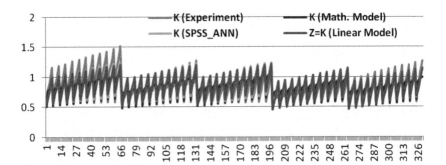

Figure 6.9 Comparison of experimental, mathematical model, ANN and linear log model values for thermal conductivity based on concentration (K$_\Phi$) (πDI).

Table 6.2 Comparison of Various Model Values for Thermal Conductivity Based on Size

(Prob-Sonication Time), K_t (π_{D2})

S.N.	K (Experiment)	K (Math. Model)	K (SPSS_ ANN)	Z = K (Linear Model)	%Error betn Expm & Math	%Error betn Expm & ANN	%Error betn Expm & SPSS
1	0.975	0.7769	0.98	0.767	20.321	0.5128	21.334
2	1.058	0.9141	1.06	0.8841	13.6	0.189	16.435
3	1.136	1.0054	1.14	1.0013	11.499	0.3521	11.861
4	1.223	1.0756	1.22	1.1184	12.053	0.2453	8.5536
5	1.014	0.809	1.01	0.8091	20.217	0.3945	20.212
6	1.103	0.9519	1.1	0.9262	13.698	0.272	16.031
7	1.192	1.047	1.19	1.0433	12.169	0.1678	12.473
8	1.28	1.1201	1.28	1.1605	12.494	0	9.3398
9	1.04	0.8419	1.04	0.8511	19.048	0	18.163
10	1.136	0.9906	1.14	0.9683	12.796	0.3521	14.767
:	:	:	:	:	:	:	:
:	:	:	:	:	:	:	:
211	1.28	1.2872	1.28	1.2697	0.5595	0	0.8039
212	1.377	1.3771	1.38	1.3868	0.0051	0.2179	0.7146
213	0.92	1.0309	0.92	1.0775	12.054	0	17.121
214	1.196	1.213	1.2	1.1946	1.4227	0.3344	0.1137
215	1.326	1.3341	1.33	1.3118	0.6124	0.3017	1.0732
216	1.42	1.4273	1.42	1.4289	0.5147	0	0.6275
217	0.966	1.068	0.97	1.1196	10.555	0.4141	15.898
218	1.254	1.2566	1.25	1.2367	0.2095	0.319	1.3796
219	1.384	1.3821	1.38	1.3538	0.1383	0.289	2.1799
220	1.463	1.4786	1.46	1.471	1.068	0.2051	0.5448
Avg	**1.1236**	**1.1186**	**1.1238**	**1.1186**	**0.4461**	**0.0202**	**0.4461**

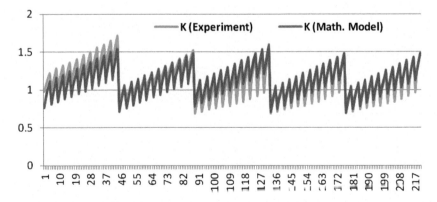

Figure 6.10 Comparison of experimental, mathematical model, ANN and linear log model values for thermal conductivity based on shape (prob-sonication time), Kt (πD2).

Table 6.3 Comparison of Various Model Values for Thermal Conductivity Based on Shape, K_s (π_{D3})

S.N.	K (Expm)	K (Math. Model)	K (SPSS_ANN)	Z = K (Linear Model)	%Error betn Expm & Math	%Error betn Expm & ANN	%Error betn Expm & SPSS
1	1.4	1.0213	1.0618	0.7853	27.053	24.157	43.91
2	1.22	1.0899	1.1101	0.8994	10.661	9.0049	26.275
3	0.99	0.9622	1.0082	1.0136	2.8125	1.8333	2.3859
4	1.44	1.0498	1.093	1.1278	27.096	24.097	21.681
5	1.25	1.1204	1.1414	0.8277	10.367	8.692	33.782
6	1.04	0.9891	1.0394	0.9419	4.8977	0.0615	9.4317
7	1.486	1.0787	1.1242	1.0561	27.409	24.346	28.931
8	1.3	1.1512	1.1726	1.1703	11.443	9.8031	9.98
9	1.08	1.0163	1.0706	0.8702	5.9008	0.8731	19.426
10	1.532	1.1079	1.1554	0.9844	27.683	24.581	35.745
⋮	⋮	⋮	⋮	⋮	⋮	⋮	⋮
⋮	⋮	⋮	⋮	⋮	⋮	⋮	⋮
156	0.946	0.9918	1.0122	1.2238	4.8372	7.0021	29.367
157	1.56	1.1174	1.1538	0.9238	28.371	26.042	40.785
158	1.28	1.1708	1.192	1.0379	8.5333	6.8734	18.913
159	0.96	1.0166	1.0434	1.1521	5.8967	8.6917	20.01
160	1.646	1.145	1.185	1.2663	30.438	28.01	23.069
161	1.333	1.1997	1.2232	0.9662	10.002	8.2348	27.516
162	1.018	1.0417	1.0747	1.0804	2.3288	5.5648	6.1287
163	1.76	1.1729	1.2162	1.1946	33.36	30.9	32.127
164	1.387	1.2289	1.2544	1.3088	11.401	9.5573	5.6417
165	1.072	1.0671	1.1059	1.0087	0.4614	3.1586	5.9067
Avg	**1.1278**	**1.104**	**1.1278**	**1.1326**	**2.11**	**5E-06**	**0.4231**

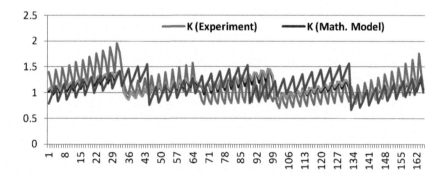

Figure 6.11 Comparison of experimental, mathematical model, ANN and linear log model values for thermal conductivity based on shape, K_s (π_{D3}).

Table 6.4 Comparison of Various Model Values for Temperature Difference, ΔT (π_{DI})

S.N.	ΔT (Expm)	ΔT (Math. Model)	ΔT (SPSS_ Linear)	ΔT (ANN)	%Error for Model with Expm	%Error for SPSS Linear with Expm	%Error for ANN with Expm
1	12.5	14.37	11.83	11.34	14.93	5.331	9.254
2	14.37	12.4	11.83	11.34	13.69	17.63	21.04
3	14.37	10.5	11.83	11.34	26.91	17.63	21.04
4	14.37	11.5	11.83	11.34	19.95	17.63	21.04
5	14.37	10.7	11.83	11.34	25.52	17.63	21.04
6	10.81	12.7	11.14	12.14	17.43	2.969	12.29
7	10.81	13.5	11.14	12.14	24.83	2.969	12.29
8	10.81	13.1	11.14	12.14	21.13	2.969	12.29
9	10.81	11.9	11.14	12.14	10.04	2.969	12.29
10	10.81	10.6	11.14	12.14	1.984	2.969	12.29
:	:	:	:	:	:	:	:
:	:	:	:	:	:	:	:
540	5.025	2.9	5.377	3.329	42.28	7.014	33.75
541	4.788	3.9	4.679	3.431	18.54	2.276	28.33
542	4.788	3.7	4.679	3.431	22.72	2.276	28.33
543	4.788	3.9	4.679	3.431	18.54	2.276	28.33
544	4.788	3	4.679	3.431	37.34	2.276	28.33
545	4.788	2.5	4.679	3.431	47.79	2.276	28.33
546	4.586	3.6	3.981	3.026	21.49	13.19	34
547	4.586	3.5	3.981	3.026	23.67	13.19	34
548	4.586	3.8	3.981	3.026	17.13	13.19	34
549	4.586	2.5	3.981	3.026	45.48	13.19	34
550	4.586	2.6	3.981	3.026	43.3	13.19	34
Avg	**6.707**	**6.946**	**6.943**	**6.968**	**3.577**	**3.527**	**3.895**

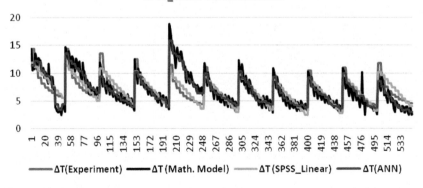

Figure 6.12 Comparison of experimental, mathematical model, ANN and linear log model values or temperature difference, ΔT (π_{DI}).

Table 6.5 Comparison of Various Model Values for Heat Flow, Q (π_{D2})

S.N.	Q (Experm)	Q (Math. Model)	Q (SPSS_ Linear)	Q(ANN)	%Error for Model with Expm	%Error for SPSS Linear with Expm	%Error for ANN with Expm
1	0.4271	0.4903	0.8593	0.4012	14.809	101.2	6.0605
2	0.4903	0.4237	0.8593	0.4012	13.596	75.247	18.178
3	0.4903	0.3588	0.8593	0.4012	26.835	75.247	18.178
4	0.4903	0.3929	0.8593	0.4012	19.867	75.247	18.178
5	0.4903	0.3656	0.8593	0.4012	25.441	75.247	18.178
6	0.7382	0.8678	0.9702	0.8519	17.557	31.424	15.399
7	0.7382	0.9225	0.9702	0.8519	24.963	31.424	15.399
8	0.7382	0.8952	0.9702	0.8519	21.26	31.424	15.399
9	0.7382	0.8132	0.9702	0.8519	10.152	31.424	15.399
10	0.7382	0.7243	0.9702	0.8519	1.8812	31.424	15.399
:	:	:	:	:	:	:	:
541	1.3963	1.1378	1.4652	0.9701	18.514	4.9273	30.526
542	1.3963	1.0795	1.4652	0.9701	22.693	4.9273	30.526
543	1.3963	1.1378	1.4652	0.9701	18.514	4.9273	30.526
544	1.3963	0.8753	1.4652	0.9701	37.319	4.9273	30.526
545	1.3963	0.7294	1.4652	0.9701	47.766	4.9273	30.526
546	1.4859	1.167	1.5761	1.0211	21.464	6.0641	31.283
547	1.4859	1.1346	1.5761	1.0211	23.646	6.0641	31.283
548	1.4859	1.2318	1.5761	1.0211	17.101	6.0641	31.283
549	1.4859	0.8104	1.5761	1.0211	45.461	6.0641	31.283
550	1.4859	0.8428	1.5761	1.0211	43.28	6.0641	31.283
Avg	1.0265	1.0634	1.0633	1.0528	3.5956	3.5844	2.5655

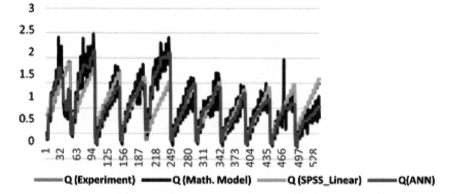

Comparision of Q values of Experimental, Mathematical Model, SPSS_Linear Model and ANN

━━ Q (Experiment) ━━ Q (Math. Model) ━━ Q (SPSS_Linear) ━━ Q(ANN)

Figure 6.13 Comparison of experimental, mathematical model, ANN and linear log model values for heat flow, Q (π_{D2}).

Table 6.6 Comparison of Various Model Values for Heat Transfer Coefficient, h (π_{D3})

S.N.	h (Experm)	h (Math. Model)	h(SPSS_ Linear)	h(ANN)	%Error for Model with Expm	%Error for SPSS Linear with Expm	%Error for ANN with Expm
1	0.049	0.064	0.107	0.053	28.68	116	8.138
2	0.064	0.051	0.107	0.053	20.14	67.88	15.96
3	0.064	0.049	0.107	0.053	22.95	67.88	15.96
4	0.064	0.055	0.107	0.053	13.12	67.88	15.96
5	0.064	0.057	0.107	0.053	9.632	67.88	15.96
6	0.094	0.098	0.12	0.107	3.971	27.63	14.27
7	0.094	0.111	0.12	0.107	17.68	27.63	14.27
8	0.094	0.112	0.12	0.107	19.34	27.63	14.27
9	0.094	0.109	0.12	0.107	16.28	27.63	14.27
10	0.094	0.106	0.12	0.107	13.16	27.63	14.27
:	:	:	:	:	:	:	:
:	:	:	:	:	:	:	:
541	0.173	0.127	0.179	0.119	26.56	3.312	31.26
542	0.173	0.126	0.179	0.119	27.4	3.312	31.26
543	0.173	0.139	0.179	0.119	19.83	3.312	31.26
544	0.173	0.116	0.179	0.119	32.93	3.312	31.26
545	0.173	0.104	0.179	0.119	39.79	3.312	31.26
546	0.184	0.131	0.192	0.125	28.78	4.532	32.18
547	0.184	0.132	0.192	0.125	28.09	4.532	32.18
548	0.184	0.15	0.192	0.125	18.51	4.532	32.18
549	0.184	0.107	0.192	0.125	41.94	4.532	32.18
550	0.184	0.119	0.192	0.125	34.99	4.532	32.18
Avg	**0.128**	**0.131**	**0.131**	**0.13**	**2.796**	**2.776**	**1.916**

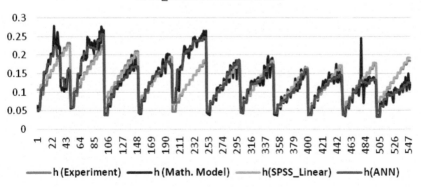

Comparision of h values of Experimental, Mathematical Model, SPSS_Linear Model and ANN

———h (Experiment) ———h (Math. Model) ———h(SPSS_Linear) ———h(ANN)

Figure 6.14 Comparison of experimental, mathematical model, ANN and linear log model values for heat transfer coefficient, h (π_{D3}).

Chapter 7

Analysis from the Indices
of the Model

7.1 INTRODUCTION

The indices of the model indicate how the phenomenon is affected due to the interaction of various independent pi terms in the models. The influence of indices of the various independent pi terms on each dependent pi term is shown in Tables 7.1 and 7.2 and discussed in this chapter.

For Two-Wire Method

Table 7.1 Constant and Indices of Response Variable

Pi Terms	$K\varphi$	K_t	K_s
K	1.51E-02	5.85E-05	1.60E+00
π_1	−0.7166	−0.7615	−1.0979
π_2	1.5642	2.476	1.6933
π_3	0.483	0.2347	0.1461

For Experimental Model Radiator as a Heat Exchanger

Table 7.2 Constant and Indices of Response Variable

Pi Terms	Temperature Difference, ΔT	Heat flow, Q	Heat Transfer Coefficient, h
K	38.1967	38.7216	30.7585
π_1	0.2028	0.1698	0.1161
π_2	−12.5534	−12.5273	−10.2355
π_3	0.1366	0.1368	0.1362
π_4	0.0457	0.046	0.0362
π_5	−0.4097	0.5903	0.5634

DOI: 10.1201/9781003432111-7

7.2 ANALYSIS OF THE MODEL FOR DEPENDENT PI TERM $\pi_{D1}\left(K_\phi\right)$

The model for the dependent pi term π_{D1} is as follows:

$$\left(K_\phi\right) = 1.51E - 02\left\{(\rho)^{-0.7166}\,(T)^{1.5642}\,(\phi)^{0.483}\right\}$$

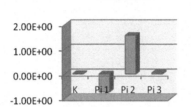

Figure 7.1 Indices of dependent pi term $\pi_{D1}\left(K_\phi\right)$.

$$\pi_{D1} = 1.51\times10^{-2}\,\left(\pi_1\right)^{-0.7166}\left(\pi_2\right)^{1.5642}\left(\pi_3\right)^{0.483} \tag{7.1}$$

The deduced equation for this pi term is given by,

$$\pi_{D1} = \left(K_\phi\right) \tag{7.2}$$

Equation (7.1) and Figure 7.1 display a model of a pi term containing thermal conductivity based on concentration K_ϕ as a response variable. The following primary conclusions drawn appear to be justified from the aforementioned model.

1. The absolute index of π_2 is highest at 1.5642. The factor π_2 is related to temperature which is the most influencing term in this model. The value of this index is positive, indicating that involvement of temperature has a strong impact on π_{D1} and that π_{D1} is directly varying with respect to π_2.
2. The absolute index of π_3 is lowest at 0.483. Thus, π_3, the term related to concentration, is the least influencing term in this model. Low value of absolute index indicates that the factor concentration needs improvement.
3. The influence of the other independent pi terms present in this model is π_1 having a negative index of –0.7166. The negative index indicates a need for improvement. The negative indices indicate that π_{D1} varies inversely with respect to π_1.
4. The constant in this model is 1.51×10^{-2}. Because this value is less than one, it has no magnification effect in the value computed from the product of the various terms of the model.

7.3 ANALYSIS OF THE MODEL FOR DEPENDENT PI TERM π_{D2} (K_t)

The model for the dependent pi term π_{D2} is as follows:

$$(K_t) = 5.85 E - 05 \left\{ (\rho)^{-0.7615} (T)^{2.476} (t)^{0.2347} \right\}$$

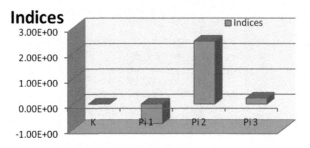

Figure 7.2 Indices of dependent pi term π_{D2} (K_t).

$$\pi_{D2} = 5.85 \times 10^{-5} (\pi_1)^{-0.7615} (\pi_2)^{2.476} (\pi_3)^{0.2347} \tag{7.3}$$

The deduced equation for this pi term is given by:

$$\pi_{D2} = (K_t) \tag{7.4}$$

Equation (7.4) and Figure 7.2 display a model of a pi term containing thermal conductivity based on size (probe sonication) K_t as a response variable. The following primary conclusion drawn appears to be justified from this model.

1. The absolute index of π_2 is highest at 2.476. The factor π_2 is related to temperature, which is the most influencing term in this model. The value of this index is positive, indicating involvement of temperature has strong impact on π_{D2} and is directly varying with respect to π_2.
2. The absolute index of π_3 is lowest at 0.2347. Thus, π_3, the term related to size (probe sonication), is the least influencing term in this model. The low value of the absolute index indicates that the factor, size (probe sonication), needs improvement.
3. The influence of the other independent pi terms present in this model is π_1 having a negative index of –0.7615. The negative index indicates a need for improvement. The negative indices indicate that π_{D2} varies inversely with respect to π_1.
4. The constant in this model is 5.85×10^{-5}. This value is less than one; hence, it has no magnification effect in the value computed from the product of the various terms of the model.

7.4 ANALYSIS OF THE MODEL FOR DEPENDENT PI TERM Π_{D3} (K_s)

The model for the dependent pi term π_{D3} is as under:

$$(K_s) = 10.5172\left\{(\rho)^{-1.7556}\,(T)^{1.6852}\,(s)^{0.0794}\right\}$$

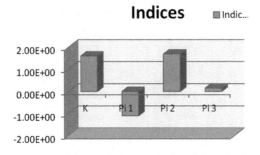

Indices   ▣ Indic...

Figure 7.3 Indices of dependent pi term π_{D3} (K_s).

$$\pi_{D3} = 10.5172(\pi_1)^{-1.7556}\,(\pi_2)^{1.6852}\,(\pi_3)^{0.0794} \tag{7.5}$$

The deduced equation for this pi term is given by:

$$\pi_{D3} = (K_s) \tag{7.6}$$

Equation (7.5) and Figure 7.3 display a model of a pi term containing thermal conductivity based on shape K_s as a response variable. The following primary conclusion drawn appears to be justified from this model.

1. The absolute index of π_2 is highest at 1.6852. The factor π_2 is related to temperature, which is the most influencing term in this model. The value of this index is positive, indicating involvement of temperature has a strong impact on π_{D1} and that π_{D1} is directly varying with respect to π_2.
2. The absolute index of π_3 is lowest at 0.0794. Thus, π_3, the term related to shape, is the least influencing term in this model. The low value of the absolute index indicates that the factor, concentration, needs improvement.
3. The influence of the other independent pi terms present in this model is π_1 having a negative index of –1.7556. The negative index indicates a need for improvement. The negative indices indicating that π_{D3} varies inversely with respect to π_1.
4. The constant in this model is 10.5172. This value is greater than one; hence, it has a magnification effect in the value computed from the product of the various terms of the model.

7.5 ANALYSIS OF THE MODEL FOR DEPENDENT PI TERM $\Pi_{D1}(\Delta T)$

The model for the dependent pi term π_{D1} is as follows:

$$(\Delta T) = 1.57\,E + 08\left\{(\varphi)^{0.2028}\,(\rho)^{-12.5534}\,(t)^{0.1366}\,(S)^{0.0457}\,(m_f)^{-0.4097}\right\}$$

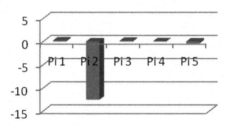

Figure 7.4 Indices of dependent pi term π_{D1} (ΔT).

$$\pi_{D1} = 1.57\,E + 08 \times (\pi_1)^{0.2028} \times (\pi_2)^{-12.5534} \times (\pi_3)^{0.1366} \times (\pi_4)^{0.0457}$$
$$\times (\pi_5)^{-0.4097} \qquad (7.7)$$

The deduced equation for this pi term is given by:

$$\pi_{D1} = (\Delta T) \qquad (7.8)$$

Equation (7.7) and Figure 7.4 display a model of a pi term containing temperature difference as a responsive variable. The following primary conclusion drawn appears to be justified from the above model.

1. The absolute index of π_1 is highest at 0.2028. The factor π_1 is related to concentration of nano fluid particles which is the most influencing term in this model. The value of this index is positive indicating involvement of temperature has strong impact on πD1 and is directly varying with respect to π_1.
2. The absolute index of π_3 is moderate viz. 0.1366. Thus π_3, the term related to size of nanofluid particles, is a moderately influencing positive term in this model.
3. The absolute index of π4 is lowest viz. 0.0457. Thus π4, the term related to shape of nano fluid particles which is the least influencing positive term in this model. The low value of the absolute index indicates that the factor, shape, needs improvement.
4. The influence of the other independent pi terms present in this model is from π_2 and π_5 having negative indices of –12.5534 and –0.4097,

respectively. The negative indices indicate a need for improvement, and they indicate that π_{D1} varies inversely with respect to π_2 and π_5.

5. The constant in this model is 1.57×10^8. This value is greater than one; hence, it has a high magnification effect in the value computed from the product of the various terms of the model.

7.6 ANALYSIS OF THE MODEL FOR DEPENDENT PI TERM Π_{D2} (Q)

The model for the dependent pi term π_{D2} is as follows:

$$(Q) = 5.27\,E + 38\left\{(\varphi)^{0.1698}\,(\rho)^{-12.5273}\,(t)^{0.1368}\,(S)^{0.046}\,(m_f)^{0.5903}\right\}$$

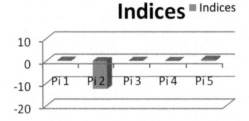

Figure 7.5 Indices of dependent pi term π_{D2} (Q).

$$\pi_{D2} = 5.27\,E + 38 \times (\pi_1)^{0.1698} \times (\pi_2)^{-12.5273} \times (\pi_3)^{0.1368} \times (\pi_4)^{0.046}$$
$$\times (\pi_5)^{0.5903} \tag{7.9}$$

The deduced equation for this pi term is given by:

$$\pi_{D2} = (Q) \tag{7.10}$$

Equation (7.9) and Figure 7.5 display a model of a pi term containing Heat Flown Q as a response variable. The following primary conclusion drawn appears to be justified from this model.

1. The absolute index of π_5 is highest viz.0.5903. The factor π_5 is related to mass flow rate of nano fluid which is the most influencing term in this model. The value of this index is positive indicating involvement of mass flow rate has strong impact on π_{D2} and is directly varying with respect to π_5.

2. The absolute indices of π_1, π_3 and π_4 is 0.1698, 0.1368 and 0.046, respectively. The value of these indices is positive, indicating involvement of concentration, size and shape of nanoparticles play an important role and is directly varying with respect to π_1, π_3 and π_4.

3. The influence of the other independent pi term present in this model is π_2, with a negative index of -12.5273. The negative index indicates a need for improvement, and it indicates that π_{D2} varies inversely with respect to π_2.

4. The constant in this model is 5.27×10^{38}. This value being greater than one, hence it has high magnification effect in the value computed from the product of the various terms of the model.

7.7 ANALYSIS OF THE MODEL FOR DEPENDENT PI TERM Π_{D3} (H)

The model for the dependent pi term π_{D3} is as follows:

$$(h) = 5.73E + 30\left\{ (\varphi)^{0.1461} (\rho)^{-10.2355} (t)^{0.1362} (S)^{0.0362} (m_f)^{0.5634} \right\}$$

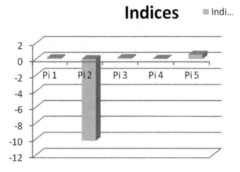

Figure 7.6 Indices of dependent pi term π_{D3} (h).

$$\pi_{D3} = 5.73E + 30 \times (\pi_1)^{0.1461} \times (\pi_2)^{-10.2355} \times (\pi_3)^{0.1362} \times (\pi_4)^{0.0362}$$
$$\times (\pi_5)^{0.5634}$$

(7.11)

The deduced equation for this pi term is given by,

$$\pi_{D3} = (h)$$

(7.12)

Equation (7.11) and Figure 7.6 display a model of a pi term containing the heat transfer coefficient h as a response variable. The following primary conclusion drawn appears to be justified from this model.

1. The absolute index of π_5 is highest at 0.5634. The factor π_5 is related to mass flow rate of nanofluid, which is the most influencing term in this

model. The value of this index is positive, indicating involvement of mass flow rate has a strong impact on π_{D3} and is directly varying with respect to π_5.

2. The absolute indices of $\pi 1$, $\pi 3$ and $\pi 4$ is 0.1461, 0.1362 and 0.0362, respectively. The values of these indices are positive in descending order, indicating involvement of concentration, size and shape of nanoparticles play an important role, and is directly varying with respect to π_1, π_3 and π_4.

3. The influence of the other independent pi term present in this model is π_2, with a negative index of -10.2355. The negative index indicates a need for improvement. The negative index indicates that π_{D3} varies inversely with respect to π_2.

4. The constant in this model is 5.73×10^{30}. This value is greater than one; hence, it has a high magnification effect in the value computed from the product of the various terms of the model.

Chapter 8

Optimization and Sensitivity Analysis

8.1 INTRODUCTION

The influences of various independent p terms have been studied by analyzing the indices of the various independent pi terms in the models. Through the technique of sensitivity analysis, the change in the value of a dependent p term caused by an introduced change in the value of an individual independent p term is evaluated. In this case, the change of ±10% is introduced in the individual p terms independently (one at a time). Thus, the total range of the introduced change is 20%. The effect of this introduced change on the percentage change value of the dependent p term is evaluated. The average value of the change in the dependent p terms is due to the introduced change of 20% in each p term. The models are in nonlinear form; hence, they are to be converted into a linear form for optimization purposes.

8.2 OPTIMIZATION OF THE MODELS

The chapter details the developed mathematical models for the two-wire method and the radiator as a heat exchanger experimental setup. The ultimate objective is to determine the best set of independent variables that will result in maximization/minimization of the objective functions. Here, the sample calculation is done for the experimental setup radiator as a heat exchanger. In this case, there are three response variables and five independent variables. The models have nonlinear form; hence, it is to be converted into a linear form for optimization purposes. This can be achieved by taking the log of both sides of the model. The linear programming technique is applied as follows:

$$\pi_{D1} = K_1 \times (\pi_1)^{a1} \times (\pi_2)^{b1} \times (\pi_3)^{c1} \times (\pi_4)^{d1} \times (\pi_5)^{e1}$$

Taking log of both the sides of the equation, we have:

$$\log \pi_{D1} = \log K_1 + a_1 \log(\pi_1) + b_1 \log(\pi_2) + c_1 \log(\pi_3) + d_1 \log(\pi_4) + e_1 \log(\pi_5)$$

DOI: 10.1201/9781003432111-8

Let

$$\log \pi_{D1} = Z, \log K_1 = k_1, \log(\pi_1) = X_1, \log(\pi_2) = X_2, \log(\pi_3)$$
$$= X_3, \log(\pi_4) = X_4, \log(\pi_5) = X_5.$$

Then the linear model in the form of first degree polynomial can be written as $Z = k + (a_1 \times X_1) + (b_1 \times X_2) + (c_1 \times X_3) + (d_1 \times X_4) + (e_1 \times X_5)$.

Thus, the equation constitutes for the optimization or, to be very specific, the maximization for the purpose of formulating the problem. The constraints can be the boundaries defined for the various independent pi terms involved in the function. During the experimentation the ranges for each independent pi terms have been defined, so that there will be two constraints for each independent variable. If one denotes maximum and minimum values of a dependent pi term π_{D1} by π_{D1max} and π_{D1min}, respectively, then the first two constraints for the problem will be obtained by taking the log of these quantities and by substituting the values of multipliers of all other variables except the one under consideration equal to zero. Let the log of the limits be defined as C_1 and C_2 {i.e., C_1 = log (π_{D1max}) and C_2 = log (π_{D1min})}. Thus, the equations of the constraints will be as follows:

$$1 \times X_1 + 0 \times X_2 + 0 \times X_3 + 0 \times X_4 + 0 \times X_5 \leq C_1$$
$$1 \times X_1 + 0 \times X_2 + 0 \times X_3 + 0 \times X_4 + 0 \times X_5 \geq C_2$$

The other constraints can be likewise found as follows:

$$0 \times X_1 + 1 \times X_2 + 0 \times X_3 + 0 \times X_4 + 0 \times X_5 \leq C_3$$
$$0 \times X_1 + 1 \times X_2 + 0 \times X_3 + 0 \times X_4 + 0 \times X_5 \geq C_4$$
$$0 \times X_1 + 0 \times X_2 + 1 \times X_3 + 0 \times X_4 + 0 \times X_5 \leq C_5$$
$$0 \times X_1 + 0 \times X_2 + 1 \times X_3 + 0 \times X_4 + 0 \times X_5 \geq C_6$$
$$0 \times X_1 + 0 \times X_2 + 0 \times X_3 + 1 \times X_4 + 0 \times X_5 \leq C_7$$
$$0 \times X_1 + 0 \times X_2 + 0 \times X_3 + 1 \times X_4 + 0 \times X_5 \geq C_8$$
$$0 \times X_1 + 0 \times X_2 + 0 \times X_3 + 0 \times X_4 + 1 \times X_5 \leq C_9$$
$$0 \times X_1 + 0 \times X_2 + 0 \times X_3 + 0 \times X_4 + 1 \times X_5 \geq C_{10}$$
$$0 \times X_1 + 0 \times X_2 + 0 \times X_3 + 0 \times X_4 + 0 \times X_5 \leq C_{11}$$
$$0 \times X_1 + 0 \times X_2 + 0 \times X_3 + 0 \times X_4 + 0 \times X_5 \geq C_{12}$$
$$0 \times X_1 + 0 \times X_2 + 0 \times X_3 + 0 \times X_4 + 0 \times X_5 \leq C_{13}$$
$$0 \times X_1 + 0 \times X_2 + 0 \times X_3 + 0 \times X_4 + 0 \times X_5 \geq C_{14}$$

After solving this linear programming problem, one obtains the minimum value of Z. The values of the independent pi terms can then be obtained by finding the antilog of the values of Z, X_1, X_2, X_3, X_4 and X_5. The actual values of the multipliers and the variables are found. The actual problem in this case can be stated as follows:

This can be solved as a linear programming problem using the MS Solver available in MS Excel.

$$\pi_{01} = K_1 \times (\pi_1)^{a1} \times (\pi_2)^{b1} \times (\pi_3)^{c1} \times (\pi_4)^{d1} \times (\pi_5)^{e1}$$

Taking log of both the sides of the equation, one has

$$\log\pi_{01} = \log K_1 + a1(\pi_1) + b1\log(\pi_2) + c1(\pi_3) + d1(\pi_4) + e1(\pi_5)$$
$$\log\pi_0 = Z, \log K1 = k1, \log(\pi_1) = X_1, \log(\pi2) = X_2, \log(\pi3) = X_3.$$

Then the linear model in the form of first-degree polynomial can be written as $Z = K + a \times X_1 + b \times X_2 + c \times X_3 + d \times X_4 + e \times X_5$.

8.2.1 For Two-Wire Method

The actual problem is to maximize Z (i.e., maximum thermal conductivity), where

$$(K_\varnothing) = 1.51E - 02\left\{(\rho)^{-0.7166}\,(T)^{1.5642}\,(\varphi)^{0.483}\right\}$$

Taking log of both the sides of the equation, we get

$$Log(K_\varnothing) = \log(1.51E - 02) - 0.7166 \times \log(\rho) + 1.5642 \times \log(T) + 0.483 \times \log(\varphi)$$

$$Z = K1 + a \times X1 + b \times X2 + c \times X3 \quad \text{and}$$

$$Z = \log(1.51E - 02) - 0.7166 \times \log(\pi_1) + 1.5642 \times \log(\pi_2) + 0.483 \times \log(\pi_3)$$

Maximum thermal conductivity based on concentration, K_φ:

$$Z1 = -1.8219 - 0.7166 \times X1 + 1.5642 \times X2 + 0.483 \times X3$$

Maximum thermal conductivity based on size, K_t:

$$Z_2 = -5.7948 - 0.2038 \times X_1 + 2.476 \times X_2 + 0.2347 \times X_3$$

Maximum thermal conductivity based on shape, K_s:

$$Z_3 = 0.2054 - 1.0979 \times X_1 + 1.6933 \times X_2 + 0.1461 \times X_3$$

Subject to the following constraints:

$$1 \times X_1 + 0 \times X_2 + 0 \times X_3 \leq 3.05038$$

Table 8.1 Optimized Values of Response Variables

	$K_{\varphi}: \pi_{D1max}$		$K_t: \pi_{D2max}$		$K_s: \pi_{D3max}$	
	Log values of π terms	Antilog of π terms	Log values of π terms	Antilog of π terms	Log values of π terms	Antilog of π terms
Z	0.027198	1.064627	0.3178432	2.0789463	0.1811372	1.517529
X_1	3.007748	1018	3.0077477	1018	3.0077477	1018
X_2	2.547775	353	2.5477747	353	2.5477747	353
X_3	0.39794	2.5	1.7781512	60	1.8404197	69.25

$$1 \times X_1 + 0 \times X_2 + 0 \times X_3 \geq 3.007748$$
$$0 \times X_1 + 1 \times X_2 + 0 \times X_3 \leq 2.547775$$
$$0 \times X_1 + 1 \times X_2 + 0 \times X_3 \geq 2.481443$$
$$0 \times X_1 + 0 \times X_2 + 1 \times X_3 \leq 0.39794$$
$$0 \times X_1 + 0 \times X_2 + 1 \times X_3 \geq -4$$

In solving this problem sing MS Solver, we get values of X_1, X_2, X_3 and Z.

8.2.2 For Experimental Model (Radiator as a Heat Exchanger)

The actual problem is to maximize Z (i.e., maximum temperature difference ΔT, heat flow Q and heat transfer coefficient h)

Maximum temperature difference, ΔT:

$$Z_1 = 38.1967 + 0.2028 \times X_1 - 12.5534 \times X_2 + 0.1366 \times X_3 + 0.0457 \times X_4 - 0.4097 \times X_5$$

Maximum heat flow, Q:

$$Z_2 = 38.7216 + 0.1698 \times X_1 - 12.5273 \times X_2 + 0.1368 \times X_3 + 0.046 \times X_4 + 0.5903 \times X_5$$

Maximum heat transfer coefficient, h:

$$Z_3 = 30.7585 + 0.1161 \times X_1 - 10.2355 \times X_2 + 0.1362 \times X_3 + 0.0362 \times X_4 + 0.5634 \times X_5$$

Subject to the following constraints:

$$1 \times X_1 + 0 \times X_2 + 0 \times X_3 + 0 \times X_4 + 0 \times X_5 \leq 0.39794$$
$$1 \times X_1 + 0 \times X_2 + 0 \times X_3 + 0 \times X_4 + 0 \times X_5 \geq -0.30103$$
$$0 \times X_1 + 1 \times X_2 + 0 \times X_3 + 0 \times X_4 + 0 \times X_5 \leq 3.044932$$
$$0 \times X_1 + 1 \times X_2 + 0 \times X_3 + 0 \times X_4 + 0 \times X_5 \geq 3.025715$$
$$0 \times X_1 + 0 \times X_2 + 1 \times X_3 + 0 \times X_4 + 0 \times X_5 \leq 3.653213$$

Table 8.2 Optimized Values of Response Variables

	$\Delta T: \pi_{D1max}$		$Q: \pi_{D2max}$		$h: \pi_{D3max}$	
	Log Values of π Terms	Antilog of π Terms	Log Values of π Terms	Antilog of π Terms	Log values of π Terms	Antilog of π Terms
Z	1.31233	20.52711	0.41273	2.5866092	−0.539172	0.2889534
X_1	0.39794	2.5	0.39794	2.5	0.39794	2.5
X_2	3.02572	1061	3.02572	1061	3.025715	1061
X_3	3.65321	4500	3.65321	4500	3.653213	4500
X_4	−7.2851	5.187E-08	−7.2851	5.187E-08	−7.285084	5.187E-08
X_5	−2.0792	0.008333	−1.0792	0.0833333	−1.079181	0.0833333

$$0 \times X_1 + 0 \times X_2 + 1 \times X_3 + 0 \times X_4 + 0 \times X_5 \geq 2.954243$$
$$0 \times X_1 + 0 \times X_2 + 0 \times X_3 + 1 \times X_4 + 0 \times X_5 \leq -7.28508$$
$$0 \times X_1 + 0 \times X_2 + 0 \times X_3 + 1 \times X_4 + 0 \times X_5 \geq -7.64016$$
$$0 \times X_1 + 0 \times X_2 + 0 \times X_3 + 0 \times X_4 + 1 \times X_5 \leq -1.07918$$
$$0 \times X_1 + 0 \times X_2 + 0 \times X_3 + 0 \times X_4 + 1 \times X_5 \geq -2.07918$$

Solving this problem using MS Solver generates values of X_1, X_2, X_3, X_4, X_5 and Z. Thus, π_{D1max} = Antilog of Z, and corresponding to this value of the π_{D1min}, the values of the independent pi terms are obtained by taking the antilog of X_1, X_2, X_3, X_4, X_5 and Z. The same procedure is adopted for π_{D2} and π_{D3}.

8.3 SENSITIVITY ANALYSIS FOR TWO-WIRE METHOD

The influence of the various independent π terms has been studied by analyzing the indices of the various π terms in the models. The technique of sensitivity analysis, the change in the value of a dependent π term caused due to an introduced change in the value of an individual π term is evaluated. In this case, change of ±10% is introduced in the individual independent π term independently (one at a time). Thus, the total range of the introduced change is ±20%. This defines sensitivity. The nature of variation in response variables due to increase in the values of independent pi terms is given in Table 8.3.

8.3.1 Effect of Introduced Change on the Dependent Pi Term π_{D1} (K_Φ)

When a total range of the change of ±10% is introduced in the value of independent pi term π_1 (ρ), a change of about 14.444% occurs in the value of π_{D1} (K_Φ) (computed from the model). The change brought in the value of π_{D1} because of change in the values of the other independent pi term π_3 (Φ) is only 0.969%. Similarly, the change of about 31.27% takes place because of change in the values of π_2 (T).

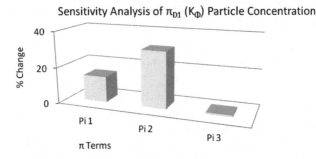

Figure 8.1 Graph of sensitivity analysis for π_{D1} (K_ϕ).

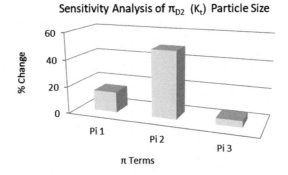

Figure 8.2 Graph of sensitivity analysis for π_{D2}.

It can be seen that the greatest amount of change takes place because of π_2, whereas the least amount of change takes place due to π_3. Thus, π_2 is the most sensitive pi term, and π_3 is the least sensitive pi term. The sequence of the various pi terms in descending order of sensitivity is π_2, π_1 and π_3.

8.3.2 Effect of Introduced Change on the Dependent Pi Term π_{D2} (K_t)

When a total range of the change of ±10% is introduced in the value of the independent pi term π_1 (ρ), a change of about 15.35459% occurs in the value of π_{D2} (K_t) (computed from the model). The change brought in the value of π_{D1} because the change in the value of the other independent pi term π_3 (t) is only 4.7046246%. Similarly the change of about 49.578009% takes place because of change in the values of π_2 (T).

It can be seen that the greatest amount of change takes place because of π_2 (T), whereas the least amount of change takes place due to π_3 (t). Thus, π_2 (T) is the most sensitive pi term and π_3 (t) is the least sensitive pi term. The sequence of the various pi terms in the descending order of sensitivity is π_2, π_1 and π_3.

8.3.3 Effect of Introduced Change on the Dependent Pi Term π_{D3} (K_s)

When a total range of the change of $\pm 10\%$ is introduced in the value of the independent pi term π_1 (ρ), a change of about 22.19835% occurs in the value of π_{D3} (K_s) (computed from the model). The change brought in the value of π_{D1} is because the change in the value of the other independent pi term π_3 (s) is only 2.9297521%. Similarly, the change of about 33.85398% takes place because of change in the values of π_2 (T).

It can be seen that the greatest amount of change takes place because of π_2 (T), whereas the least amount of change takes place due to π_3 (s). Thus, π_2 (T) is the most sensitive pi term, and π_3 (s) is the least sensitive pi term. The sequence of the various pi terms in the descending order of sensitivity is π_2, π_1 and π_3.

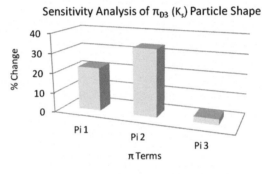

Sensitivity Analysis of π_{D3} (K_s) Particle Shape

Figure 8.3 Graph of sensitivity analysis for π_{D3} (K_s).

Table 8.3 Sensitivity Analysis for Two-Wire Method

π_1 (ρ)	π_2 (T)	π_3 (Φ, t, s)	π_{D1} (K_Φ)	π_{D2} (K_t)	π_{D3} (K_s)
1081	328	1.25001667	0.8791671	1.1365028	1.1076384
1189.1	328	1.25001667	0.8211253	1.0569391	0.997592
972.9	328	1.25001667	0.9481155	1.2314445	1.2434695
		% Change	14.444376	15.35459	22.198355
1081	328	1.25001667	0.8791671	1.1365028	1.1076384
1081	360.8	1.25001667	1.0205114	1.4389934	1.301632
1081	295.2	1.25001667	0.7455856	0.8755379	0.9266523
		% Change	31.271157	49.578009	33.85398
1081	328	1.25001667	0.8791671	1.1365028	1.1076384
1081	328	1.37501833	0.8832236	1.162212	1.1231699
1081	328	1.125015	0.8747044	1.1087439	1.0907188
		% Change	0.969008	4.7046246	2.9297521

8.3.4 Effect of Introduced Change on the Dependent Pi Term π_{D1} (ΔT)

When a total range of the change of $\pm 10\%$ is introduced in the value of independent pi terms π_1 (Φ), π_3 (t) and π_4 (s), a change of about 4.07%, 2.74% and 0.92% occurs in the value of π_{D1} (ΔT) (computed from the model). The change brought in the value of π_{D1} because of change in the values of the other independent pi terms π_2 (ρ) and π_5 (m_f) is 345% and 8.24%, respectively. Table 8.4 shows that the greatest amount of change takes place because of the pi term π_2, whereas the least amount of change takes place due to the pi term π_4. Thus, π_2 is the most sensitive pi term and π_4 is the least sensitive pi term.

8.3.5 Effect of Introduced Change on the Dependent Pi Term π_{D2} (Q)

When a total range of the change of $\pm 10\%$ is introduced in the value of independent pi terms π_1 (Φ), π_3 (t) and π_4 (s), a change of about 3.405%, 2.743% and 0.923% occurs in the value of π_{D2} (Q) (computed from the model). The change brought in the value of π_{D1} because of change in the values of the other independent pi terms π_2 (ρ) and π_5 (m_f) is 344% and 11.82%, respectively. It can be seen that the greatest amount of change takes place because of the pi term π_2, whereas the least amount of change takes place due to the pi term π_4. Thus, π_2 is the most sensitive pi term and π_4 is the least sensitive pi term.

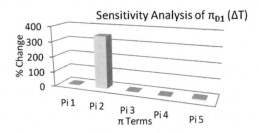

Figure 8.4 Graph of sensitivity analysis for π_{D1} (ΔT).

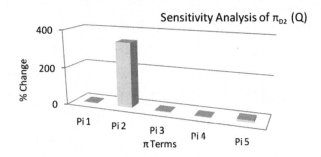

Figure 8.5 Graph of sensitivity analysis for π_{D2} (Q).

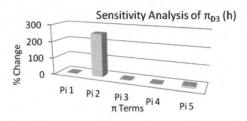

Figure 8.6 Graph of sensitivity analysis for π_{D3} (h).

Table 8.4 Sensitivity Analysis for Radiator as a Heat Exchanger Experimental Setup

	±10% Change in Independent Terms					π_{D1}	π_{D2}	π_{D3}
	π_1 (Φ)	π_2 (ρ)	π_3 (t)	π_4 (s)	π_5 (m_f)	ΔT	Q	h
Avg.	2.0455	1098.09	3190.9	4.8E-08	0.046	6.06	1.086	0.13
10%	2.25	1098.09	3190.9	4.8E-08	0.046	6.17	1.104	0.14
-10%	1.8409	1098.09	3190.9	4.8E-08	0.046	5.93	1.067	0.13
				%Change		4.07	3.405	2.33
Avg.	2.0455	1098.09	3190.9	4.8E-08	0.046	6.06	1.086	0.13
10%	2.0455	1207.9	3190.9	4.8E-08	0.046	1.83	0.329	0.05
-10%	2.0455	988.282	3190.9	4.8E-08	0.046	22.7	4.066	0.4
				%Change		345	344	256
Avg.	2.0455	1098.09	3190.9	4.8E-08	0.046	6.06	1.086	0.13
10%	2.0455	1098.09	3510	4.8E-08	0.046	6.14	1.101	0.14
-10%	2.0455	1098.09	2871.8	4.8E-08	0.046	5.97	1.071	0.13
				%Change		2.74	2.743	2.73
Avg.	2.0455	1098.09	3190.9	4.8E-08	0.046	6.06	1.086	0.13
10%	2.0455	1098.09	3190.9	5.3E-08	0.046	6.08	1.091	0.14
-10%	2.0455	1098.09	3190.9	4.4E-08	0.046	6.03	1.081	0.13
				%Change		0.92	0.923	0.73
Avg.	2.0455	1098.09	3190.9	4.8E-08	0.046	6.06	1.086	0.13
10%	2.0455	1098.09	3190.9	4.8E-08	0.05	5.82	1.149	0.14
-10%	2.0455	1098.09	3190.9	4.8E-08	0.041	6.32	1.021	0.13
				%Change		8.24	11.82	11.3

8.3.6 Effect of Introduced Change on the Dependent Pi Term π_{D3} (h)

When a total range of the change of ±10% is introduced in the value of independent pi terms π_1 (Φ), π_3 (t) and π_4 (s), a change of about 2.33%, 2.73% and 0.73% occurs in the value of π_{D3} (h) (computed from the model). The change brought in the value of π_{D1} because of change in the values of the other independent pi terms π_2 (ρ) and $\pi5$ (m_f) is 256% and 11.3%, respectively. It can be seen that greatest amount of change takes place because of the pi term π_2, whereas the least amount of change takes place due to the pi term π_4. Thus, π_2 is the most sensitive pi term and π_4 is the least sensitive pi term.

8.4 ESTIMATION OF LIMITING VALUES OF RESPONSE VARIABLES

The ultimate objective of this work is to find out best set of variables, which will result in maximization/minimization of the response variables. This section outlines an attempt to find out the limiting values of three response variables for the two-wire method viz. thermal conductivity based on concentration ($K_\emptyset$), size (K_t) and shape (K_s). To achieve this, limiting values of independent π terms π_1 (ρ), π_2 (T) and π_3 (Φ) are put in the respective models. In the process of maximization, the maximum value of the independent π term is substituted in the model if the index of the term is positive, and the minimum value is inserted if the index of the term is negative. The limiting values of these response variables computed for the two-wire method are in Table 8.5.

8.4.1 For Two-Wire Method

$$() = 1.51 - 02\left\{()^{-0.7166}\,()^{1.5642}\,()^{0.483}\right\}$$

$$(K_t) = 5.85E - 05\left\{(\rho)^{-0.7615}\,(T)^{2.476}\,(t)^{0.2347}\right\}$$

$$(K_s) = 10.5172\left\{(\rho)^{-1.7556}\,(T)^{1.6852}\,(s)^{0.0794}\right\}$$

8.4.2 For Radiator as a Heat Exchanger Experimental Setup

This section outlines an attempt to find out the limiting values of three response variables for the radiator as a heat exchanger experimental setup (empirical model) temperature difference ΔT, total heat flow (Q) and heat transfer coefficient (h). To achieve this, limiting values of independent π terms π_1 (Φ), π_2 (ρ), π_3 (t), π_4 (s) and π_5 (m_f) are put in the respective models. In the process of

Table 8.5 Maxima and Minima for Two-Wire Method

	K_1					Unit
MAX	0.01507	1018	353	2.5	1.064627399	W/m°K
MIN	0.01507	1123	303	0.0001	0.479148824	W/m°K
	K_1					Unit
MAX	5.85E-05	1018	353	60	1.593390913	W/m°K
MIN	5.85E-05	6310	303	15	0.196543285	W/m°K
	K_1					Unit
MAX	1.604723	1018	353	5.187E-08	1.422295667	W/m°K
MIN	1.604723	1175	303	2.29E-08	0.832538245	W/m°K

Table 8.6 Maxima and Minima for Radiator as a Heat Exchanger

	Max. Temperature Difference ΔT °K						π_{D1}	
	π_1	π_2	π_3	π_4	π_5	K	ΔT	
MAX	2.5	1061	4500	5.2E-08	0.008	2E+38	20.5	°K
MIN	0.5	1109	900	2.3E-08	0.083	2E+38	4.46	°K
	Total Heat Q (Watt)						π_{D2}	
	π_1	π_2	π_3	π_4	π_5	K	Q	
MAX	2.5	1061	4500	5.2E-08	0.083	5E+38	2.59	Watt
MIN	0.5	1109	900	2.3E-08	0.008	5E+38	0.22	Watt
	Overall Heat Transfer Coefficient, h W/m² °K						$\pi D3$	
	π_1	π_2	π_3	π_4	π_5	K	h	
MAX	2.5	1061	4500	5.2E-08	0.083	6E+30	0.29	W/m²°K
MIN	0.5	1109	900	2.3E-08	0.008	6E+30	0.03	W/m²°K

maximization, the maximum value of the independent π term is substituted in the model if the index of the term is positive, and the minimum value is inserted if the index of the term is negative. The limiting values of these response variables are computed for radiator as a heat exchanger are as given in Table 8.6.

$$(\Delta T) = 1.57\,E+08\left\{(\varphi)^{0.2028}(\rho)^{-12.5534}(t)^{0.1366}(S)^{0.0457}(m_f)^{-0.4097}\right\}$$

$$(Q) = 5.27\,E+38\left\{(\varphi)^{0.1698}(\rho)^{-12.5273}(t)^{0.1368}(S)^{0.046}(m_f)^{0.5903}\right\}$$

$$(h) = 5.73\,E+30\left\{(\varphi)^{0.1461}(\rho)^{-10.2355}(t)^{0.1362}(S)^{0.0362}(m_f)^{0.5634}\right\}$$

8.5 PERFORMANCE OF THE MODELS

The models have been formulated mathematically as well as by using the ANN. Now we have three sets of values of three dependent pi terms of heat transfer performance values computed by experimental observations, values computed by mathematical model and values obtained by ANN. All these values match very well with each other. This is justified by calculating their respective mean values and standard errors of estimation.

From the values of "standard error estimation" and "percent error" shown Figure 8.7, it seems that the mathematical models and the ANN developed using MATLAB® can be successfully used for computation of dependent terms for a given set of the independent pi terms.

The values of R^2 error in ANN, number of iterations, values of the regression coefficients for dependent pi terms and the plots of the actual data and target data for the dependent pi terms are shown in Figures 8.7, 8.8 and 8.9. The values are fairly good.

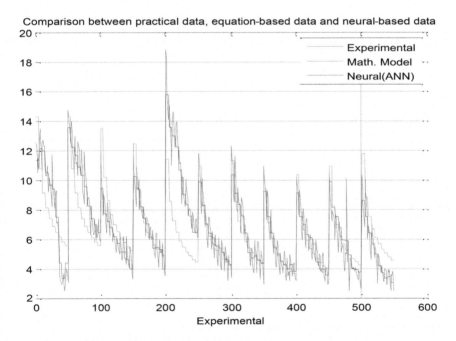

Figure 8.7 Neural network model, epoches, training, validation, R value and comparison of results of experimental, model and ANN for π_{d1} (temperature difference, ΔT).

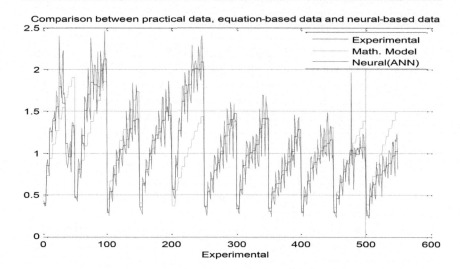

Figure 8.8 Neural network model, epoches, training, validation, R value and comparison of results of experimental, model and ANN for π_{D2} (heat flow, Q).

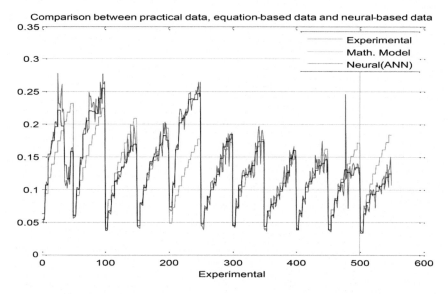

Figure 8.9 Neural network model, epoches, training, validation, R value and comparison of results of experimental, model and ANN for π_{D3} (heat transfer coefficient, h).

8.6 RELIABILITY OF MODELS

This section presents an approach to the reliability of models. Error frequency distribution graphs were plotted for the developed models. These graphs were compared with probability density function graphs of commonly used life distributions.

In statistical analysis, frequency distribution is the most generalized case. Many statistical distributions are used to model various reliability parameters. The particular distribution used depends on the nature of the data being analyzed. Model reliability approximation is executed by comparing error frequency graphs of various mathematical models with probability density function graphs of commonly used life distribution.

This section also presents a statistical method that explains how much of the variability of a factor can be caused or explained by its relationship to another factor. This is achieved through the trend analysis by the method of coefficient of determination. The approach of R^2 (coefficient of determinant) is used to analyze the behavior of mathematical models. The calculation for the reliability and the value of R^2 is done for three response variables, i.e., π_{D1} to π_{D3} and the proper comparisons are made to analyze the best suit for the reliability and the value of R^2.

8.6.1 Life Distribution

Life distribution is the basic tool of the reliability engineer, which may also be called a failure distribution. They can be either a combination of smaller distributions of different failure mechanism or a single distribution representing single failure mechanism. Life distribution can have any shape, but some standard forms have become commonly used over the years. Commonly used life distributions are: normal, lognormal, exponential and Weibull.

8.6.1.1 The Normal Distribution

The normal distribution is more often used to model repair, although it is also used to model reliability. In this application, the normal distribution is most applicable to simple maintenance tasks that consistently require a fixed amount of time to complete with little variation. The probability density function of the normal distribution is often called the bell curve because of its distinctive shape. Figure 8.10 shows the plot of the standard normal probability density function.

8.6.1.2 The Lognormal Distribution

The lognormal distribution is also used to model reliability. It applies to maintenance tasks in which the task time and frequency vary, which is often the case for complex systems and products. Unlike the mean of the normal

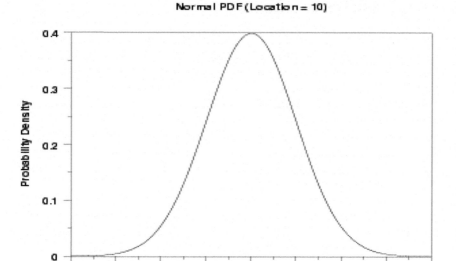

Figure 8.10 Probability density function normal distribution.

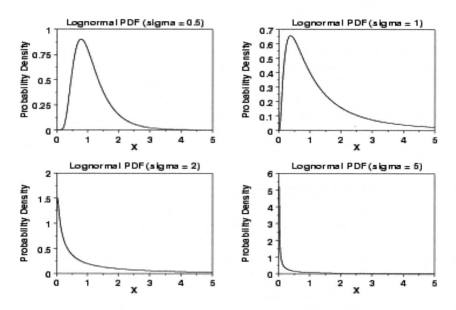

Figure 8.11 Probability density function lognormal distribution.

distribution, the mean of the lognormal is not the 50th percentile of the distribution, and the distribution is not symmetrical around the mean. Figure 8.11 shows the plot of the standard lognormal probability density function for four values of σ.

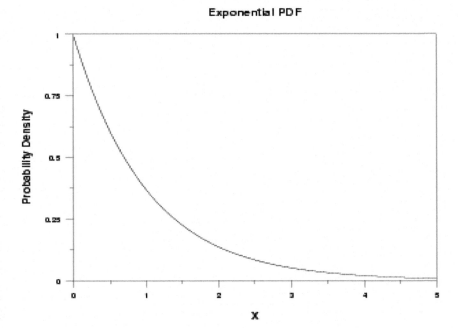

Figure 8.12 Probability density function exponential distribution.

8.6.1.3 The Exponential Distribution

The exponential distribution is widely used to model electronic reliability failures in the operating domain that tend to exhibit a constant failure rate. To fail exponentially means the distribution of failure times fits the exponential distribution plot, as shown in Figure 8.12 with regard to probability density function.

8.6.1.4 The Exponential Distribution

The Weibull distribution is an important distribution, since it can be used to represent many different probability density functions. It has many applications and can be used to fit a wide variety of data. Figure 8.13 illustrates the plot of the Weibull probability density function.

8.6.2 Reliability Analysis

A subject that is so important to many decisions in this world could hardly escape quantitative analysis. The name "reliability" is given to the field of study that attempts to assign numbers to the propensity of systems to fail. In a more restrictive sense, the term "reliability" is defined as probability of a system to perform its mission successfully. This section presents an approach

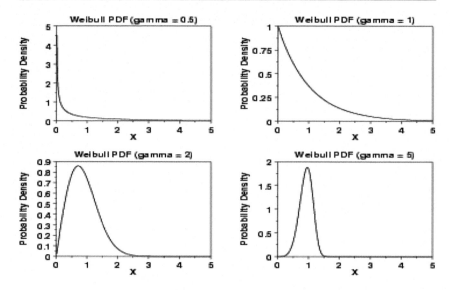

Figure 8.13 Probability density function weibull distribution.

to reliability of models, using error frequency distribution for developed models with graphical representation. These graphs were compared with probability density function graphs of commonly used life distributions.

Reliability of the model can be established by using Reliability % = 100 – % Mean Error

$$\% \, of \, Mean \, Error = \frac{\sum_i x_i f_i}{\sum_i f_i} \qquad (8.1)$$

where x_i is the % Mean Error, and f_i is the frequency Mean Error

8.6.3 Error Frequency Distribution

In our experimentation, we have two sets of values – i.e., observed and computed values – of all three dependent parameters of (1) the two-wire method, and (2) radiator as a heat exchanger experimental setup. For the two-wire method, thermal conductivity based on concentration, K_Φ (π_{D1}), thermal conductivity based on size, K_t (π_{D2}), and thermal conductivity based on shape, K_s (π_{D3}), are measured. For radiator as a heat exchanger experimental setup, temperature difference, ΔT (π_{D1}) heat flow, Q (π_{D2}), hear transfer coefficient, h (π_{D3}). The difference of these two values provides the error. Frequencies of occurrence of specific errors are estimated for Two Wire Method and Radiator as a Heat Exchanger Experimental Setup. Tables 8.7 to 8.11 show calculations for frequency of error analysis and Reliability. Figures 8.14 to 8.18 show graphs of % Error Frequency for all response variable models.

8.6.3.1 For Two-Wire Method

Table 8.7 Error Frequency Distribution and Reliability for π_{DI} (Thermal Conductivity Based on Concentration, K_Φ)

Error	Freq.	% Error (f_i)	Frequency (x_i)	$f_i * x_i$
0	15	0	15	0
1	14	1	14	14
2	12	2	12	24
3	15	3	15	45
4	11	4	11	44
5	9	5	9	45
6	10	6	10	60
7	21	7	21	147
8	25	8	25	200
9	25	9	25	225
10	12	10	12	120
11	11	11	11	121
12	12	12	12	144
13	6	13	6	78
14	15	14	15	210
15	9	15	9	135
16	14	16	14	224
17	7	17	7	119
18	5	18	5	90
19	4	19	4	76
20	5	20	5	100
21	9	21	9	189
22	8	22	8	176
23	8	23	8	184
24	6	24	6	144
25	3	25	3	75
26	2	26	2	52
27	2	27	2	54
28	3	28	3	84
29	7	29	7	203
30	5	30	5	150
31	4	31	4	124
33	4	33	4	132
34	3	34	3	102
35	3	35	3	105
36	1	36	1	36
37	1	37	1	37
38	2	38	2	76
39	2	39	2	78
		748	**330**	**4222**
		Mean Error	$\Sigma (f_i * x_i)/x_i$	12.793939
		Reliability	**(100-Mean Error)**	87.206061

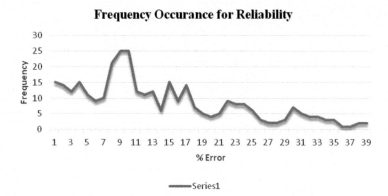

Figure 8.14 Percentage error frequency graph for K_Φ (π_{D1}).

Table 8.8 Error Frequency Distribution and Reliability for π_{D2} (Thermal Conductivity Based on Size)

% Error (f_i)	Frequency (x_i)	$f_i * x_i$
0	24	0
1	11	11
2	24	48
3	20	60
4	9	36
5	3	15
6	9	54
7	9	63
8	10	80
9	10	90
10	19	190
11	18	198
12	12	144
13	8	104
14	6	84
15	9	135
16	7	112
17	1	17
18	2	36
19	2	38
20	3	60
21	2	42
22	1	22
23	1	23
276	**220**	**1662**
Mean Error	$\Sigma (f_i * x_i)/x_i$	7.5545455
Reliability	(100-Mean Error)	92.445455

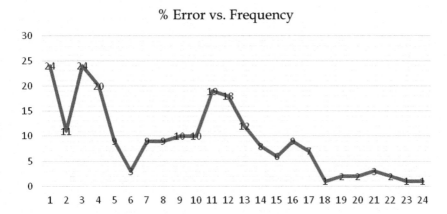

Figure 8.15 Percentage error frequency graph for π_{D2}.

Table 8.9 Error Frequency Distribution and Reliability for π_{D3} (Thermal Conductivity Based on Size)

% Error (f_i)	Frequency (x_i)	$f_i * x_i$
0	4	0
1	3	3
2	5	10
3	4	12
4	2	8
5	1	5
6	3	18
7	3	21
8	2	16
9	1	9
10	3	30
11	1	11
12	1	12
13	3	39
14	1	14
15	3	45
16	7	112
17	4	68
18	2	36
19	4	76
20	3	60
21	2	42
22	1	22
23	2	46

(Continued)

Table 8.9 (Continued)

% Error (f_i)	Frequency (x_i)	$f_i * x_i$
24	2	48
25	4	100
26	2	52
27	2	54
28	1	28
29	1	29
30	2	60
31	1	31
35	1	35
36	1	36
40	1	40
43	1	43
45	1	45
46	1	46
48	1	48
49	1	49
57	1	57
63	1	63
958	**90**	**1579**
Mean Error	$\Sigma\ (f_i * x_i)/x_i$	17.544444
Reliability	**(100-Mean Error)**	82.455556

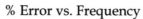

% Error vs. Frequency

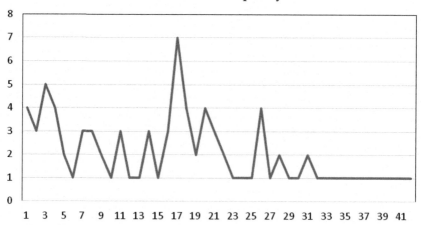

Figure 8.16 Percentage error vs. frequency graph for π_{D3}.

8.6.3.2 For Radiator as a Heat Exchanger Experimental Setup

Table 8.10 Error Frequency Distribution and Reliability for π_{D1} (Temperature Difference ΔT)

% Error (f_i)	Frequency (x_i)	$f_i * x_i$
0	15	0
1	18	18
2	14	28
3	17	51
4	9	36
5	14	70
6	21	126
7	15	105
8	17	136
9	14	126
10	15	150
11	19	209
12	13	156
13	18	234
14	22	308
15	16	240
16	16	256
17	14	238
18	13	234
19	4	76
20	19	380
21	7	147
22	11	242
23	16	368
24	9	216
25	11	275
26	11	286
27	11	297
28	2	56
29	8	232
30	10	300
31	2	62
32	8	256
33	5	165
34	7	238
35	6	210
36	7	252
37	7	259
38	4	152
39	13	507
40	5	200
41	4	164
42	5	210
43	3	129
44	1	44
45	1	45
46	5	230

(Continued)

Table 8.10 (Continued)

% Error (f_i)	Frequency (x_i)	f_i * x_i
47	6	282
48	1	48
49	2	98
50	2	100
51	1	51
52	4	208
53	2	106
54	1	54
57	2	114
59	3	177
61	1	61
63	1	63
64	1	64
65	1	65
69	1	69
71	2	142
73	3	219
76	2	152
78	1	78
80	1	80
83	1	83
91	2	182
94	2	188
97	2	194
100	3	300
2766	**550**	**11897**
Mean Error	$\Sigma\ (f_i * x_i)/x_i$	**21.630909**
Reliability	**(100-Mean Error)**	**78.369091**

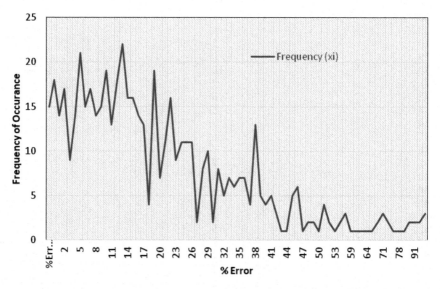

Figure 8.17 Percentage error vs. frequency graph for $\Delta T\ (\pi_{D1})$.

Table 8.11 Error Frequency Distribution and Reliability for π_{D2} (Heat Flow, Q)

% Error (f_i)	Frequency (x_i)	$f_i * x_i$
0	18	0
1	15	15
2	13	26
3	16	48
4	10	40
5	15	75
6	22	132
7	17	119
8	14	112
9	14	126
10	15	150
11	19	209
12	12	144
13	19	247
14	24	336
15	15	225
16	15	240
17	14	238
18	12	216
19	7	133
20	16	320
21	9	189
22	12	264
23	14	322
24	11	264
25	10	250
26	10	260
27	11	297
28	2	56
29	9	261
30	10	300
31	2	62
32	7	224
33	5	165
34	7	238
35	7	245
36	6	216
37	7	259
38	4	152
39	13	507
40	5	200
41	5	205
42	4	168
43	3	129
44	1	44
45	1	45
46	5	230
47	6	282
48	1	48
49	2	98

(Continued)

Table 8.11 (Continued)

% Error (f_i)	Frequency (x_i)	$f_i * x_i$
50	2	100
51	2	102
52	4	208
53	2	106
54	1	54
56	1	56
57	1	57
59	3	177
61	1	61
63	1	63
64	1	64
65	1	65
69	1	69
71	2	142
73	3	219
76	2	152
78	1	78
80	1	80
83	1	83
91	2	182
94	2	188
97	2	194
100	3	300
2822	551	11931
Mean Error	$\Sigma (f_i * x_i)/x_i$	**21.653358**
Reliability	**(100-Mean Error)**	**78.346642**

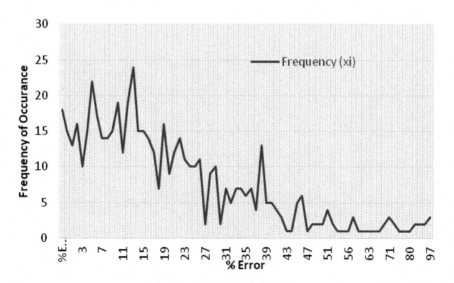

Figure 8.18 Percentage error frequency graph for Q (π_{D2}).

Table 8.12 Error Frequency Distribution and Reliability for π_{D3} (Heat Transfer Coefficient, h)

% Error (f_i)	Frequency (x_i)	$f_i * x_i$
0	153	0
1	181	181
2	139	278
3	59	177
4	13	52
5	5	25
15	550	713
Mean Error	$\Sigma (f_i * x_i)/x_i$	1.2963636
Reliability	**(100-Mean Error)**	98.703636

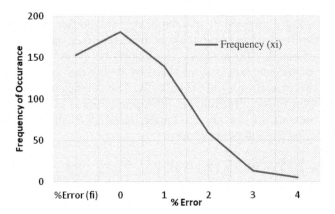

Figure 8.19 Percentage error vs. frequency graph for h (π_{D3}).

For two-wire method empirical model reliability:

For π_{D1}, (thermal conductivity based on concentration, K_{ϕ}, is 87.20%.
For π_{D2}, thermal conductivity based on size, K_r, is 92.44%.
For π_{D3}, thermal conductivity based on shape, K_s, is 82.45%.

For experimental empirical model using radiator as a heat exchanger

Reliability:

For π_{D1}, temperature difference, ΔT, is 78.36%
For π_{D2}, Heat flow, Q) is 81.93%.
For π_{D3}, (Heat transfer coefficient, h) is 78.34%.

From calculation, the value of Reliability R_i for general mathematical models is greater than 78%, so the model is more accurate.

8.7 COEFFICIENT OF DETERMINANT R^2 FOR TWO-WIRE METHOD

A statistical method that explains how much of the variability of a factor can be caused or explained by its relationship to another factor, coefficient of determinant is used in trend analysis. It is computed as a value between 0 (0 percent) and 1 (100 percent). The higher the value, the better the fit. Coefficient of determinant is symbolized by R^2 because it is square of the coefficient of correlation symbolized by R. The coefficient of determinant is an important tool in determining the degree of linear correlation of variables (goodness of fit) in regression analysis and also called R-square. It is calculated using the following relation:

$$R^2 = 1 - \sum(Yi - fi)^2 / \sum(Yi - Y)^2 \tag{8.2}$$

where

   Yi = Observed value of dependent variable for i-th experimental sets
   (experimental data)
   fi = Observed value of dependent variable for i-th predicted value sets
   (model data)
   Y = Mean of Yi and
   R^2 = Coefficient of determinant

From calculation the value of R^2 for general models are nearer to 1 and combined models are nearer to zero. A value of general model indicates a nearly perfect fit and is therefore a reliable model for future forecasts.

8.7.1 For Two-Wire Method

8.7.1.1 Coefficient of Determinant (R^2) for π_{DI} (Thermal Conductivity Based on Concentration, K_ϕ)

Table 8.13 Sample Calculation of Coefficient of Determinant (R^2) for π_{DI} (Thermal Conductivity Based on Concentration, K_ϕ)

F_i	Y_i				
$Z_I = K$ (Model)	$Z_I = K$ (Expm)	$(Y_i - f_i)$	$(Y_i - f_i)^2$	$(Y_i - Y)$	$(Y_i - Y)^2$
0.50695	0.5612	0.054249523	0.002943	−0.28018	0.0785
0.764937	0.772	0.007062791	4.99E-05	−0.06938	0.004813
0.79098	0.85	0.059019939	0.003483	0.008622	7.43E-05
0.806623	0.924	0.117376735	0.013777	0.082622	0.006826
0.81791	1.049	0.231090439	0.053403	0.207622	0.043107
0.826773	1.16	0.33322747	0.111041	0.318622	0.10152
0.520097	0.5637	0.043603409	0.001901	−0.27768	0.077105
0.784773	0.798	0.013226628	0.000175	−0.04338	0.001882
0.811492	0.876	0.06450844	0.004161	0.034622	0.001199
0.82754	0.963	0.135459581	0.018349	0.121622	0.014792
:	:	:	:	:	:
:	:	:	:	:	:
:	:	:	:	:	:
0.918094	0.898	−0.020094344	0.000404	0.056622	0.003206
0.949352	0.96	0.010648456	0.000113	0.118622	0.014071
0.968127	1.084	0.115873142	0.013427	0.242622	0.058865
0.981673	1.16	0.178327083	0.031801	0.318622	0.10152
0.99231	1.267	0.274689554	0.075454	0.425622	0.181154
			$\Sigma(Y_i - f_i)^2$		$\Sigma(Y_i - Y)^2$
273.5667	277.6547	4.087977121	**7.321179**	−9E-05	**15.3133**
	Y				
0.82899	**0.841378**	0.012387809	0.022185	−2.7E-07	0.046404
			$X = \Sigma(Y_i - f_i)^2/$		**0.478093**
			$\Sigma(Y_i - Y)^2$		
				$R^2 = 1 - x$	**0.521907**
				R	**0.722431**

8.7.1.2 Coefficient of Determinant (R^2) for π_{D2} (Thermal Conductivity Based on Size, K_t)

Table 8.14 Sample Calculation of Coefficient of Determinant (R^2) for π_{D2} (Thermal Conductivity based on Size, K_t)

fi	Yi				
Z = K (Model)	Z = K (Expm)	(Yi – fi)	(Yi – fi)²	(Yi – Y)	(Yi – Y)²
0.77687	0.975	0.19813	0.03926	−0.0997	0.00995
0.91411	1.058	0.14389	0.02070	−0.0167	0.00028
1.00537	1.136	0.13063	0.01706	0.06126	0.00375
1.07560	1.223	0.14740	0.02173	0.14826	0.02198
0.80900	1.014	0.20500	0.04203	−0.0607	0.00369
0.95191	1.103	0.15109	0.02283	0.02826	0.0008
1.04695	1.192	0.14505	0.02104	0.11726	0.01375
1.12008	1.280	0.15992	0.02557	0.20526	0.04213
0.84190	1.04	0.19810	0.03924	−0.0347	0.00121
0.99064	1.136	0.14536	0.02113	0.06126	0.00375
1.08954	1.231	0.14146	0.02001	0.15626	0.02442
1.03090	0.920	−0.11090	0.01230	−0.1547	0.02394
:	:	:	:	:	:
1.06796	0.966	−0.102	0.0104	−0.1087	0.01182
1.25663	1.254	−0.0026	6.9E-06	0.17926	0.03213
1.38209	1.384	0.00191	3.7E-06	0.30926	0.09564
1.47863	1.463	−0.0156	0.00024	0.38826	0.15075
			Σ(Yi – fi)²		Σ(Yi –Y)²
246.087	247.19	1.1028	2.45713	10.7474	11.0299
	Y				
1.06994	1.07474	0.00479	0.01068	0.04673	0.04796
			x = Σ(Yi – fi)²/ Σ(Yi –Y)²		0.22277
				R² =1 – x	0.77723
				R	0.88161

8.7.1.3 Coefficient of Determinant (R^2) for π_{D3} (Thermal Conductivity Based on Shape, K_s)

Table 8.15 Sample Calculation of Coefficient of Determinant (R^2) for π_{D3} (Thermal Conductivity Based on Shape, K_s)

fi	Yi				
$Z = K$ (Model)	$Z = K$ (Expm)	$(Yi - fi)$	$(Yi - fi)^2$	$(Yi - Y)$	$(Yi - Y)^2$
1.0213	1.4	0.3787	0.1434	0.5909	0.3492
1.0899	1.22	0.1301	0.0169	0.4109	0.1688
0.9622	0.99	0.0278	0.0008	0.1809	0.0327
1.0498	1.44	0.3902	0.1522	0.6309	0.398
1.1204	1.25	0.1296	0.0168	0.4409	0.1944
0.9891	1.04	0.0509	0.0026	0.2309	0.0533
1.0787	1.486	0.4073	0.1659	0.6769	0.4582
1.1512	1.3	0.1488	0.0221	0.4909	0.241
1.0163	1.08	0.0637	0.0041	0.2709	0.0734
1.1079	1.532	0.4241	0.1799	0.7229	0.5226
:	:	:	:	:	:
1.1708	1.28	0.1092	0.0119	0.4709	0.2218
1.0166	0.96	−0.057	0.0032	0.1509	0.0228
1.145	1.646	0.501	0.251	0.8369	0.7004
1.1997	1.333	0.1333	0.0178	0.5239	0.2745
1.0417	1.018	−0.024	0.0006	0.2089	0.0436
1.1729	1.76	0.5871	0.3447	0.9509	0.9042
1.2289	1.387	0.1581	0.025	0.5779	0.334
1.0671	1.072	0.0049	2E-05	0.2629	0.0691
			$\Sigma(Yi - fi)^2$		$\Sigma(Yi - Y)^2$
182.17	186.09	3.9266	9.2697	52.591	29.219
	Y				
0.792	0.8091	0.0171	0.0403	0.2287	0.127
			$x = \Sigma(Yi - fi)^2 / \Sigma(Yi - Y)^2$		0.3173
				$R^2 = 1 - x$	0.6827
				R	0.8263

8.7.2 For Radiator as a Heat Exchanger Experimental Setup

8.7.2.1 Coefficient of Determinant (R^2) for π_{D1} (Temperature Difference, ΔT)

Table 8.16 Sample Calculation for π_{D1} (Temperature Difference, ΔT)

fi	Yi	(Yi – fi)	(Yi – fi)²	(Yi – Y)	(Yi –Y)²
ΔT (Model)	ΔT (Expm)				
14.3661	12.5	–1.8660961	3.482315	5.556909	30.87924
14.3661	12.4	–1.9660961	3.865534	5.456909	29.77786
14.3661	10.5	–3.8660961	14.9467	3.556909	12.6516
14.3661	11.5	–2.8660961	8.214507	4.556909	20.76542
:	:	:	:	:	:
fi	**Yi**	**(Yi – fi)**	**(Yi – fi)²**	**(Yi – Y)**	**(Yi – Y)²**
ΔT (Model)	ΔT (Expm)				
:	:	:	:	:	:
4.585636	3.5	–1.085636449	1.178606	–3.44309	11.85488
4.585636	3.8	–0.785636449	0.617225	–3.14309	9.879021
4.585636	2.5	– 2.085636449	4.349879	–4.44309	19.74106
4.585636	2.6	–1.985636449	3.942752	–4.34309	18.86244
			Σ(Yi – fi)²		Σ(Yi –Y)²
3690.477	3818.7	128.2225572	**2311.941**	–5E-05	**4789.189**
	Y				
6.709959	6.943091	0.233131922	4.20353	–9.1E-08	8.707616
			x = Σ(Yi – fi)² / Σ(Yi –Y)²		0.482742
			R2=1 – x		0.517258
			R		0.719207

8.7.2.2 Coefficient of Determinant (R^2) for π_{D2} (Heat Flow, Q)

Table 8.17 Sample Calculation for π_{D2} (Heat Flow, Q)

fi	Yi	(Yi – fi)	(Yi – fi)²	(Yi – Y)	(Yi – Y)²
Q (Model)	Q (Expm)				
0.49033	0.42708	–0.0632	0.004	–0.6362	0.40471
0.49033	0.42367	–0.0667	0.00444	–0.6396	0.40907
0.49033	0.35875	–0.1316	0.01731	–0.7045	0.49633

(Continued)

fi	Yi	(Yi – fi)	(Yi – fi)²	(Yi – Y)	(Yi – Y)²
Q (Model)	Q (Expm)				
0.49033	0.39292	−0.0974	0.00949	−0.6703	0.44935
:	:	:	:	:	:
:	:	:	:	:	:
:	:	:	:	:	:
1.48595	1.13458	−0.3514	0.12346	0.07133	0.00509
1.48595	1.23183	−0.2541	0.06458	0.16858	0.02842
1.48595	0.81042	−0.6755	0.45635	−0.2528	0.06393
1.48595	0.84283	−0.6431	0.4136	−0.2204	0.04859
			$\Sigma(Yi - fi)^2$		$\Sigma(Yi - Y)^2$
564.617	584.79	20.1727	**60.6499**	−0.0003	**132.222**
	Y				
1.02658	1.06325	0.03668	0.11027	−5E-07	0.2404
			$x = \Sigma(Yi - fi)^2 / \Sigma(Yi - Y)^2$		**0.4587**
			$R^2 = 1 - x$		**0.5413**
			R		**0.73573**

8.7.2.3 Coefficient of Determinant (R^2) for π_{D3} (Heat Transfer Coefficient, h)

Table 8.18 Sample Calculation for π_{D3} (Heat Transfer Coefficient, h)

fi	Yi	(Yi – fi)	(Yi – fi)²	(Yi – Y)	(Yi – Y)²
h (Model)	h (Expm)				
0.0635	0.0494	−0.014	0.0002	−0.082	0.0067
0.0635	0.0507	−0.013	0.0002	−0.081	0.0065
0.0635	0.049	−0.015	0.0002	−0.082	0.0068
0.0635	0.0552	−0.008	7E-05	−0.076	0.0058
0.0635	0.0574	−0.006	4E-05	−0.074	0.0055
:	:	:	:	:	:
:	:	:	:	:	:
:	:	:	:	:	:
0.1837	0.1321	−0.052	0.0027	0.0006	4E-07
0.1837	0.1497	−0.034	0.0012	0.0182	0.0003
0.1837	0.1067	−0.077	0.0059	−0.025	0.0006

(Continued)

Table 8.18 (Continued)

fi	Yi	(Yi − fi)	(Yi − fi)²	(Yi − Y)	(Yi − Y)²
h (Model)	h (Expm)				
0.1837	0.1194	−0.064	0.0041	−0.012	0.0001
			Σ(Yi − fi)²		Σ(Yi −Y)²
70.365	72.303	1.9386	**0.7495**	0.0002	**1.7573**
	Y				
0.1279	0.1315	0.0035	0.0014	4E-07	0.0032
			x = Σ(Yi − fi)² / Σ(Yi −Y)²		**0.4265**
			R² = 1 − x		**0.5735**
			R		**0.7573**

8.7.3 Coefficient of Determinant Results

For the two-wire method empirical model:

Coefficient of Determinant R^2:

For π_{D1}, thermal conductivity based on concentration, K_ϕ, is 0.722431.
For π_{D2}, thermal conductivity based on size, K_s, is 0.777231.
For π_{D3}, thermal conductivity based on shape, K_s, is 0.682.

For radiator as a heat exchanger experimental empirical model:

Coefficient of Determinant R^2

For π_{D1}, temperature difference, ΔT, is 0.517258.
For π_{D2}, heat flow, Q, is 0.541303.
For π_{D3}, heat transfer coefficient, h, is 0.573471.

From calculation, the value of R^2 for general mathematical models are shown in Tables 8.13 to 8.18. These tables show that the values of general model are closer, so the model is more accurate.

This section has presented the approach to reliability of mathematical models. Standard life distribution and their probability density functions are the basis for reliability approximation. A comparison of error frequency graphs with probability density function graph reveals that reliability of models is equal to reliability of these standard distributions.

Approach to the R^2 values of mathematical models were also presented. From the comparison of percentage reliability of mathematical models and of the R^2 values of mathematical models, it is noted that both approaches are found to be better.

Chapter 9

Interpretation of the Simulation

9.1 INTERPRETATION OF INDEPENDENT VARIABLES VS. RESPONSE VARIABLES AFTER OPTIMIZATION

After following the three approaches, it is observed that the results converge to the same conclusion as established by the two-wire (hot) method. That is, Al_2O_3/W nanofluid has the highest thermal conductivity, followed by TiO_2, ZrO_2, MgO and CuO. Also, the concentration of nanoparticles has the highest impact on increasing the thermal conductivity followed by size and shape of nanoparticles.

1. Based on the two-wire method, a test was developed for three response variables in terms of independent variables. Indices of independent variables indicate that increase in the thermal conductivity of nanofluid is highest because of concentration, moderate because of size and lowest because of shape of nanoparticles. This validates the results received by the two-wire method, i.e., "Al2O3 has highest thermal conductivity at 2.5 wt.% concentration, 60 min probe sonication time and Cubic shape at 353 K temp".
2. Table 9.1 very clearly narrates the values of three independent parameters derived from the optimized equation as 2.5 wt% concentration, 60 min. probe-sonication time for size and cubic shape of Al_2O_3 nanoparticles. This is exactly in agreement with the conclusion derived by the two-wire (hot) method of experimentation, which says that "Al_2O_3 has highest thermal conductivity at 2.5 wt.% concentration, 60 min probe sonication and Cubic shape at 353 K temp".
3. Table 9.2 shows the comparison of reliability of the multiple regression mathematical model, the ANN model using SPSS and the log-log linear model using SPSS. It was found that the reliability of all the response variables is excellent. Therefore, it can be stated that the reliability of response variables K_φ, K_t and K_s is perfectly satisfactory.
4. Table 9.2 also shows the coefficient of determinant R^2 of the multiple regression mathematical model, the ANN model using SPSS and the

Table 9.1 Values of Independent Variables vs. Response Variables after Optimization

S.N.	Independent Variable	Property	Response Variables		
			π_{D1} (K_ϕ)	π_{D2} (K_t)	π_{D3} (K_s)
1	π1	Density	1061	1061	1061
2	π2	Temp. (K)	353	353	353
3	π3	Φ, t, shape	2.5 wt%	60 min	Cubic

Table 9.2 Reliability (%) and Coefficient of Determinant R^2 for Response Variables

S.N.	Response Variable	Reliability %	Coefficient of Determinant R^2		
			Mathematical	SPSS	ANN
1	π_{D1} (K_ϕ)	87.20	0.722431	0.976	0.690
2	π_{D2} (K_t)	92.44	0.777231	0.988	0.212
3	π_{D3} (K_s)	82.45	0.682000	0.988	0.792

log-log linear model using SPSS. It is observed that for all the response variables, values of R^2 are the best match for the data set. Hence, it can be said that these models are suitable for the experimental optimization of heat transfer process in a heat exchanger using nanofluids.

The ANN performance depends on the training of model. The comparative lower value of the regression coefficient for one of the dependent pi terms may be due to the improper training of the network. The ANN has been unable to predict beyond the range for which it has been trained.

5. Table 9.3 explains that based on these values of mean error, percentage error between the experimental model and the mathematical model, between the experimental model and the ANN (using SPSS) model, and between the experimental model and the log-log linear (using SPSS) model, it is evident that results of percentage error values are well within the permissible limit of 15%. Hence, this also validates the present research experimental optimization of heat transfer process in a heat exchanger using nanofluids.

6. The optimized equation for radiator as a heat exchanger received for three response variables in terms of independent variables also indicate that increase in the thermal conductivity of nanofluid is highest because of concentration, moderate because of size and lowest because of shape of nanoparticles. This validates the results received by the two-wire method, i.e., "Al_2O_3 has highest thermal conductivity at 2.5 wt.% concentration, 60 min probe sonication and Cubic shape at 353 K temp".

Table 9.3 Values of Mean Error and % Error for Various Models vs. Response Variables

		Mean Error				Percentage Error Between		
S.N.	Response Variable	Experimental Model	Mathematical Model	ANN Model (SPSS)	Log-Log Linear Model (SPSS)	Expm & Math	Expm & ANN	Expm & Log Linear
1	K_Φ	0.841378	0.82899	0.837727	0.841378	1.4723	0.433866	3.60×10^{-6}
2	K_t	1.12783	1.104033	1.12783	1.132602	2.1100	5.37×10^{-6}	0.42306
3	K_s	1.12359	1.118578	1.123818	1.118578	0.4461	0.020227	0.446126

Table 9.4 Values of Independent Variables vs. Response Variables after Optimization

S.N.	Independent Variable	Property	Response Variables		
			π_{D1} (ΔT)	π_{D2} (Q)	π_{D3} (h)
1	π_1	Φ	2.5 wt%	2.5 wt%	2.5 wt%
2	π_2	P	1061	1061	1061
3	π_3	t (size)	3600	3600	3600
4	π_4	S (shape)	Cubic	Cubic	Cubic
5	π_5	mf	0.008333	0.08333	0.08333

7. Table 9.4 very clearly narrates the values of five independent parameters derived from optimized equation. It can be distinctly seen that the optimized value of concentration is 2.5 wt%, 60 min. probesonication time for size and cubic shape of Al_2O_3 nanoparticles. This is exactly in agreement with the conclusion derived by the two-wire (hot) method of experimentation, which says that "Al_2O_3 has highest thermal conductivity at 2.5 wt.% concentration, 60 min probe sonication and Cubic shape at 353 K temp".

9.2 INTERPRETATION OF TEMPERATURE DIFFERENCE AGAINST THE MASS FLOW RATE

Table 9.5 depicts the average values of inlet and outlet (radiator) temperature difference against the mass flow rate (0.5 to 5 lit/min) for water, coolant and nanofluid. The temperature difference ΔT values are maximum for nanofluid followed by coolant and water. The highest value of ΔT is observed at mass flow rate of 0.5 lit/min and lowest at 5 lit/min in all the three cases.

These two statements conclude that the highest heat transfer (thermal conductivity) is because of nanofluid followed by coolant and water.

Similarly, other graphs and tables also reveal that "Al_2O_3 has highest thermal conductivity at 2.5 wt. % concentration, 60 min probe sonication and Cubic shape at 353 K temp".

9.3 INTERPRETATION OF RELIABILITY AND COEFFICIENT OF DETERMINANT

Table 9.6 shows that in comparing the mathematical model, the ANN model using SPSS and the log-log linear model using SPSS, the reliability of all the response variables is excellent. Therefore, it can be stated that the reliability of response variables (ΔT), (Q) and (h) is perfectly satisfactory.

Also, in observing the coefficients of the determinant R^2 of the Mathematical model, ANN model using SPSS and log-log linear model using SPSS, for all the response variables, values of R^2 are the best match for the data set. Hence, it can be said that these models are suitable for the experimental optimization of the heat transfer process in a heat exchanger using nanofluids.

Table 9.5 Water, Coolant and Nanofluid Average ΔT vs. Discharge

Sr. No.	Discharge	Average of Difference of Temperature		
		Water	Coolant	Nanofluid
1	0.5	8.56	9.45	16.52
2	1	8.2	8.62	13.26
3	1.5	7.14	8.52	13.5
4	2	6.66	6.82	12.9
5	2.5	5.78	5.88	10.4
6	3	4.94	6.04	9.26
7	3.5	5.15	5.62	9.04
8	4	4.34	4.6	7.58
9	4.5	4.38	4.42	7.1
10	5	3.9	4.26	6.62

Table 9.6 Reliability (%) and Coefficient of Determinant R^2 for Response Variables

S.N.	Response Variable	Reliability %	Coefficient of Determinant R^2		
			Mathematical	SPSS	ANN
1	$\pi_{D1}(\Delta T)$	78.36	0.517258	0.992	0.943913
2	$\pi_{D2}(Q)$	81.93	0.541303	0.984	0.946843
3	π_{D3} (h)	78.34	0.573471	0.968	0.910362

9.4 INTERPRETATION OF MEAN ERROR OF MODELS CORRESPONDING TO RESPONSE VARIABLES

Tables 9.7 and 9.8 explain the mean error calculated for the Experimental model, the Mathematical model, the ANN model using SPSS and the log-log linear model using SPSS. Based on these values of mean error, percentage error between the Experimental and Mathematical models, Experimental and ANN (using SPSS) models, and Experimental and log-log linear (using SPSS) models are derived. It is evident from the percentage error that the values are well within the permissible limit of 15%. Hence, this also validates the present research experimental optimization of heat transfer process in a heat exchanger using nanofluids.

Table 9.7 Values of Mean Error of Models Corresponding to Response Variables

		Mean Error			
S.N.	Response Variables	Experimental Model	Mathematical Model	ANN model (using SPSS)	Log-log linear Model (SPSS)
1	$\pi_{D_1}(\Delta T)$	6.7066	6.946484	6.967813	6.943091
2	$\pi_{D_2}(Q)$	1.0265	1.06337	1.052795	1.063254
3	$\pi_{D_3}(h)$	0.1279	0.131486	0.130666	0.131461

Table 9.8 Values of Percentage Error of Models Corresponding to Response Variables

		Percentage Error Between		
S.N.	Response Variables	Expm & Math	Expm & ANN	Expm & Log Linear
1	$\pi_{D_1}(\Delta T)$	1.4723	0.433866	3.60×10^{-6}
2	$\pi_{D_2}(Q)$	2.1100	5.37×10^{-6}	0.42306
3	$\pi_{D_3}(h)$	0.4461	0.020227	0.446126

References

1. Fei Duan, Dingtian Kwek, Alexandru Crivoi, Viscosity affected by nanoparticle aggregation in Al_2O_3-water nanofluids, *Nanoscale Research Letters*, Vol. 6, pp. 248–253 (2011).
2. www.nano.gov/
3. Report on European Activities in the Field of Ethical, Legal and Social Aspects (ELSA) and Governance of Nanotechnology, http://cordis.europa.eu/nanotechnology/
4. Yury Gogotsi, Ed., *Nanomaterials Handbook*. Taylor & Francis Group LLC, 2006.
5. Guozhong Cao, *Nanostructures & Nanomaterials: Synthesis, Properties & Applications*. Imperial College Press, 2004.
6. P. J. Harris, *Carbon Nanotubes and Related Structures*. Cambridge University Press, 1999.
7. P. Harrison, *Quantum Wells, Wires and Dots*. John Wiley & Sons. Ltd., 2005.
8. M. F. Abdul Jalal, N. H. Shuaib, P. Gunnasegaran, E. Sandhita, Investigation on the cooling performance of a compact heat exchanger using nanofluids, *AIP Proceedings*, Vol. 520 (2012).
9. Dustin R. Ray, Debendra K. Das, Ravikanth S. Vajjha, Experimental and numerical investigations of nanofluids performance in a compact minichannel plate heat exchanger, *International Journal of Heat and Mass Transfer*, Vol. 71, pp. 732–746 (April 2014).
10. A. K. Rasheed, M. Khalid, W. Rashmi, T. C. S. M. Gupta, A. Chand, Graphene based nanofluids and nanolubricants – review of recent developments, *Renewable and Sustainable Energy Reviews*, Vol. 63, pp. 346–362 (2016).
11. Layth Ismael, Khalid Faisal Sultan, A comparative study on the thermal conductivity of micro and nano fluids by using silver and zirconium oxide, *Al-Qadisiya Journal for Engineering Sciences*, Vol. 7, pp. 189–204 (2014).
12. Veeranna Sridhara, Lakshmi Narayan Satapathy, Al2O3-based nanofluids: A review, *Nanoscale Research Letters*, Vol. 6, pp. 456–471 (2011).
13. Xiang-Qi Wang, Arun S. Mujumdar, A review on nanofluids – part II: Experiments and applications, *Brazilian Journal of Chemical Engineering*, Vol. 25, pp. 631–648 (2008).
14. S. Nabi, E. Shirani, Simultaneous effects of Brownian motion and clustering of nanoparticles on thermal conductivity of nanofluids, IJST, *Transactions of Mechanical Engineering*, Vol. 36, No. M1, pp. 53–68 (2009).

15. Clement Kleinstreuer, Yu Feng, Experimental and theoretical studies of nanofluid thermal conductivity enhancement: A review, *Nanoscale Letters*, Vol. 6, pp. 229–242 (2011).
16. E. K. Goharshadi, H. Ahmadzadeh, S. Samiee, M. Hadadian, Nanofluids for heat transfer enhancement – a review, *Physical Chemistry Research*, Vol. 1, pp. 1–33 (2013).
17. Emmanuel C. Nsofor, Tushar Gadge, Investigations on the nanolayer heat transfer in nanoparticles-in-liquid suspensions, *ARPN Journal of Engineering and Applied Sciences*, Vol. 6, pp. 21–28 (2011).
18. Y. Liu, D. Kai, Z. Xiao-Song, Influence factors on thermal conductivity of ammonia-water nanofluids, *Journal of Central South University*, Vol. 19, pp. 1622–1628 (2012).
19. Tae-Keun Hong, Ho-Soon Yang, Nanoparticle-dispersion-dependent thermal conductivity in nanofluids, *Journal of Korean Society*, Vol. 47, pp. S321–S324 (2005).
20. Elena V. Timofeeva, Jules L. Routbort, Dileep Singh, Particle shape effects on thermophysical properties of alumina nanofluids, *Journal of Applied Physics*, Vol. 106, pp. 014304–014314 (2009).
21. Kaufui V. Wong, Michael J. Castillo, Heat transfer mechanisms and clustering in nanofluids, *Advances in Mechanical Engineering*, Vol. 2010, pp. 795478–795487 (2010).
22. D. Cabaleiro, L. Colla, F. Agresti, L. Lugo, L. Fedele, Transport properties and heat transfer coefficients of ZnO/(ethylene glycol + water) nanofluids, *International Journal of Heat and Mass Transfer*, Vol. 89, pp. 433–443 (2015).
23. Gabriela Huminic, Angel Huminic, Application of nanofluids in heat exchangers: A review, *Renewable and Sustainable Energy Reviews*, Vol. 16, pp. 5625–5638 (2012).
24. S. K. Mohammadian, Y. Zhang, Analysis of nanofluid effects on thermoelectric cooling by micro-pin-fin heat exchangers, *Applied Thermal Engineering*, Vol. 70, pp. 282–290 (2014).
25. J. Albadr, S. Tayal, M. Alasadi, Heat transfer through heat exchanger using Al2O3 nanofluid at different concentrations, *Case Studies in Thermal Engineering*, Vol. 1, pp. 38–44 (2013).
26. I. M. Shahrul, I. M. Mahbubul, R. Saidur, M. F. M. Sabri, Experimental investigation on Al2O3–W, SiO2–W and ZnO–W nanofluids and their application in a shell and tube heat exchanger, *International Journal of Heat and Mass Transfer*, Vol. 97, pp. 547–558 (2016).
27. Z. Wu, L. Wang, B. Sunden, L. Wadso, Aqueous carbon nanotube nanofluids and their thermal performance in a helical heat exchanger, *Applied Thermal Engineering*, Vol. 96, pp. 364–371 (2016).
28. Dan Huang, Zan Wu, Bengt Sunden, Pressure drop and convective heat transfer of Al2O3/water and MWCNT/water nanofluids in a chevron plate heat exchanger, *International Journal of Heat and Mass Transfer*, Vol. 89, pp. 620–626 (2015).
29. Arun Kumar Tiwari, Pradyumna Ghosh, Jahar Sarkar, Particle concentration levels of various nanofluids in plate heat exchanger for best performance, *International Journal of Heat and Mass Transfer*, Vol. 89, pp. 1110–1118 (2015).

30. M. A. Khairul, A. Hossain, R. Saidur, M. A. Alim, Prediction of heat transfer performance of CuO/water nanofluids flow in spirally corrugated helically coiled heat exchanger using fuzzy logic technique, *Computers and Fluids*, Vol. 100, pp. 123–129 (2014).
31. Faroogh Garoosi, Leila Jahanshaloo, Mohammad Mehdi Rashidi, Arash Badakhsh, Mohammed E. Ali, Numerical simulation of natural convection of the nanofluid in heat exchangers using a Buongiorno model, *Applied Mathematics and Computation*, Vol. 254, pp. 183–203 (2015).
32. Wael I. A. Aly, Numerical study on turbulent heat transfer and pressure drop of nanofluid in coiled tube-in-tube heat exchangers, *Energy Conversion and Management*, Vol. 79, pp. 304–316 (2014).
33. C. J. Ho, Y. N. Chung, Chi-Ming Lai, Thermal performance of Al2O3/water nanofluid in a natural circulation loop with a mini-channel heat sink and heat source, *Energy Conversion and Management*, Vol. 87, pp. 848–858 (2014).
34. T. Srinivas, A. Venu Vinod, Heat transfer intensification in a shell and helical coil heat exchanger using water-based nanofluids, *Chemical Engineering and Processing*, Vol. 102, pp. 1–8 (2016).
35. Azher M. Abed, M. A. Alghoul, K. Sopiana, H. A. Mohammedc, Hasan Sh. Majdi, Ali Najah Al-hamani, Design characteristics of corrugated trapezoidal plate heat exchangers using nanofluids, *Chemical Engineering and Processing*, Vol. 87, pp. 88–103 (2015).
36. M. Sheikholeslami, M. Hatamia, M. Jafaryarb, F. Farkhadnia, D. D. Ganji, M. Gorji-Bandpya, Thermal management of double-pipe air to water heat exchanger, *Energy and Buildings*, Vol. 88, pp. 361–366 (2015).
37. Marjan Goodarzi, Ahmad Amiri, Mohammad Shahab Goodarzi, Mohammad Reza Safaei, Arash Karimipour, Ehsan Mohseni Languri, Mahidzal Dahari, Investigation of heat transfer and pressure drop of a counter flow corrugated plate heat exchanger using MWCNT based nanofluids, *International Communications in Heat and Mass Transfer*, Vol. 66, pp. 172–179 (2015).
38. Gabriela Huminic, Angel Huminic, Heat transfer and entropy generation analyses of nanofluids in helically coiled tube-in-tube heat exchangers, *International Communications in Heat and Mass Transfer*, Vol. 71, pp. 118–125 (2016).
39. Tun-Ping Teng, Yu-Chun Hsu, Wei-Ping Wang, Yan-Bo Fang, Performance assessment of an air-cooled heat exchanger for multiwalled carbon nanotubes- water nanofluids, *Applied Thermal Engineering*, Vol. 89, pp. 346–355 (2015).
40. Brett Baker, Mike Johns, Einar Fridjonsson, Mark Titley, Nanofluid application in LNG process heat exchange systems, *CEED Seminar Proceedings*, pp. 7–12 (2014).
41. X. J. Wang, X. F. Li, Y. H. Xu, D. S. Zhu, Thermal energy storage characteristics of Cu-H2O nanofluids, *Energy*, Vol. 78, pp. 212–217 (2014).
42. Nizar Ahammed, Lazarus Godson Asirvatham, Somchai Wongwises, Thermoelectric cooling of electronic devices with nanofluid in a multiport minichannel heat exchanger, *Experimental Thermal and Fluid Science*, Vol. 74, pp. 81–90 (2016).

43. Mushtaq I. Hasan, Abdul Muhsin A. Rageb, Mahmmod Yaghoubi, Investigation of a counter flow microchannel heat exchanger performance with using nanofluid as a coolant, *Journal of Electronics Cooling and Thermal Control*, Vol. 2, pp. 35–43 (2012).
44. Alpesh V. Mehta, Nimit M. Patel, Dinesh K. Tantia, Nilsh M. Jha, Mini heat exchanger using Al_2O_3-water based nano fluid, *International Journal of Mechanical Engineering and Technology*, Vol. 4, pp. 238–244 (2013).
45. E. Ebrahimnia-Bajestan, M. C. Moghadam, H. Niazmand, W. Daungthongsuk, S. Wongwises, Experimental and numerical investigation of nanofluids heat transfer characteristics for application in solar heat exchangers, *International Journal of Heat and Mass Transfer*, Vol. 92, pp. 1041–1052 (2016).
46. Fatou Toutie Ndoye, Patrick Schalbart, Denis Leducq, Graciela Alvarez, Numerical study of energy performance of nanofluids used in secondary loops of refrigeration systems, *International Journal of Refrigeration*, Vol. 52, pp. 122–132 (2015).
47. Singiresu S. Rao, *Engineering Optimization*, 3rd ed. New Age International (P) Limited Publishers, 2002.
48. Miller Irwin, Miller Marylees, *John E. Freund's Mathematical Statistics with Applications*, 7th ed. Pearson Education, pp. 54–55 (2006).

Index